THE LAST APE

*Pygmy Chimpanzee Behavior
and Ecology*

THE LAST APE

Pygmy Chimpanzee Behavior and Ecology

TAKAYOSHI KANO

Translated by
Evelyn Ono Vineberg

Stanford University Press
Stanford, California

Stanford University Press
Stanford, California

© 1986 Dobutsushsha
English translation © 1992 by

the Board of Trustees of the Leland
Stanford Junior University

Originally published in Japanese in 1986
by Dobutsusha as *Saigo no ruijinen*

First published in the U.S.A. by
Stanford University Press, 1992

Printed in the United States of America

Original printing 1992
Last figure below indicates year of this printing:
03 02 01 00 99 98 97 96 95 94

Stanford University Press publications are distributed exclusively by Stanford
University Press within the United States, Canada, and Mexico; they are distributed
exclusively by Cambridge University Press throughout the rest of the world.

Author's Acknowledgments

The foundation of this book is the theory developed by my esteemed Professor Junichiro Itani of Kyoto University. From several of his works, I learned to see chimpanzee society from the point of view of coexistence, rather than competition or selection. My sensei's invitation to join his research team, set up in 1973 (Research into the Ecology of Wild Chimpanzees and Primitive Hunter-Gatherer People in the African Forest-Open Land Boundary Zone), was the start of my pygmy chimpanzee research.

It was my long-time friend Professor Toshisada Nishida of Tokyo University who encouraged me to investigate pygmy chimpanzees. Not only did he suggest the course of research, he also accompanied me in the initial phase, in 1973. Later, he visited me at Wamba and offered pertinent advice and information.

Since 1974, I have been responsible for directing Japanese research on the pygmy chimpanzees, under grants from Mombusho, Japan's Ministry of Education, Science, and Culture (Grant-in-Aid for Overseas Scientific Research). The support of many professors, including Shiro Kondo and Masao Kawai, has helped me to realize and sustain my research plan.

I received precious data and offers of information from Hidemi Ishida, Ryu Asato, Suehisa Kuroda, Akio Mori, Takeshi Furuichi, Yukio Takahata, Hiroyuki Takasaki, Toshikazu Hasegawa and Mariko Hiraiwa Hasegawa, Kohji Kitamura, Shigeo Uehara, Juichi Yamagiwa, and the African research team. I benefited greatly from discussions with all of them.

Permission for local research was granted by the successive directors of IRSAC (L'Institut de Recherche Scientifique de l'Afrique Central) and CRSN (Centre de Recherche en Sciences Naturelles). I could always depend on the cooperation of the Japanese Embassy in Zaire and several businesses in Kinshasa (for machine parts, for receiving and sending materials, and for negotiations with various agencies). The employees of Japanese companies in Zaire also went to considerable trouble for me.

If I had not had the devoted service and cooperation of the African employees at the Wamba research site, I could not have carried out this research. Many people on the plantations and in the Catholic churches also went out of their way to help us.

The additional workload during my absences inconvenienced many of my colleagues at the University of the Ryukyus. I would not have been able to continue my long-term research without their deep understanding and warm support.

Professor Adrienne L. Zihlman of the University of California, Santa Cruz, gladly agreed to the reprinting of several figures from her articles. Professor Junichiro Itani and Dr. Michael Huffman kindly lent me photographs of common chimpanzees. Professor Osamu Takenaka taught me a great deal about molecular evolution, and Professor Fukuichi Nakata, at the University of the Ryukyus, advised me on blood-type terminology. I kept Mrs. Sanae Nitta busy arranging data, drawing figures and tables, and seeing to other things.

My friend from college days, Professor Kosei Izawa, took the trouble to introduce me to Mr. Ryoichi Hisaki of Dobutsusha and saw to the realization of my plans for this book. Mr. Hisaki, from start to finish, went along with me. He was patient and encouraging in allowing me to write as I pleased.

I want to offer my heartfelt appreciation to all of these people and organizations, and I deeply apologize for the omission of others.

T. K.

Author's Preface to the English Edition

Pygmy chimpanzees inhabit only the tropical forest region in the central part of the Republic of Zaire, south of the Zaire (Congo) River. They are extremely rare and valuable apes. Because they cannot be seen in zoos in Japan, they had been practically unknown there until my cherished friend Suehisa Kuroda, who worked at Wamba with me for many years, presented this enchanting ape in his splendid book *Pygmy Chimpanzee: The Unknown Ape* (*Pigumi chinpanji: Michi no ruijinen* [Tokyo: Chikuma-shobo, 1982]).

My own book was written for high school and university students and for laypeople interested in animals, with the goal of conveying details about the natural history of pygmy chimpanzees. Many results of my observations are presented in the form of tables and figures, and I also include photographs of my subjects. Because there *is* no research on pygmy chimpanzees in Japan outside of this field study, the descriptions in the first chapter of the general character and features of primates may seem like "preaching to God" to English-speaking readers, who have access to the more diverse Western literature. From the second chapter on, however, this book is based mainly on the results of my observations.

At present, there are only two long-term pygmy chimpanzee field-research sites: our Wamba Forest site, which continues to be supported by a Grant-in-Aid for Overseas Scientific Survey from the Ministry of Education, Science, and Culture, Japan, and the Lomako Forest site of the research team under the leadership of Randall Susman of the State University of New York at Stony Brook. Both field-research sites were established at about the same time, in 1973, but the Japanese team used provisioning as a method of habituating the chimpanzees, whereas the Lomako team observed only under natural conditions. Thus the data collected at the two research sites can be contrasted.

Japanese field researchers have used the provisioning method for many years, but Western field researchers seldom have. The research site for wild common chimpanzees at the celebrated Gombe National Park, Tanzania, where Jane Goodall used provisioning to aid in observations, is an exception.

In Japan, there have been many debates about the distorting influence of provisioning as a field-research method for studying primates. When wild primates are habitually provisioned, their entire ecology, beginning with their food habits and ranging patterns, is affected. But most primates quickly lose their wariness of humans when provisioned, and the observers can see them when they are at ease, no longer bothered by the presence of humans. By contrast, in the forest, where most primates live, visibility is poor, and the observer can see only some of the members of the group and only some of the activities occurring within the group. By provisioning, a researcher can induce his or her subjects to come to an open place where visibility is good (i.e., the feeding site), and can place himself or herself within the field of vision of several members of a group. The method has its shortcomings, but it also has notable advantages. Most of the observations about social relationships among pygmy chimpanzees reported in this book were obtained at the feeding site.

Social animals show a variety of general behavioral patterns, but in primates the principle of "competition" seems to explain how most individuals and groups subsist and leave descendants. In many species of primates, dominant-subordinate rank, brought out through competition, determines relationships between individuals. Once rank is established between individuals, they can live together harmoniously within a group. Among Wamba's pygmy chimpanzees, however, this formula does not particularly apply. Dominant-subordinate relationships certainly exist among pygmy chimpanzee group members, but by frequent use of quasi-sexual behavior during interactions, they conceal the operation of rank and associate with one another very adeptly. From the point of view of individual survival, they are the most successful species among the higher primates, including common chimpanzees (which are a sibling species). They prove that individuals can coexist without relying on competition and dominant-subordinate rank, and this is what I want to emphasize in this book.

The pygmy chimpanzees of Wamba's study groups have been given their own individual names. (Many of them are Japanese names from an earlier era.) However, I avoided giving names of personal acquaintances except when celebrating cases of direct observations. I used common adult male or juvenile names. I did this in part out of fear that the descriptions might become too anthropomorphic, and in part because I thought it was important to standardize rigorously, even though the research was progressing slowly.

Like other higher primates, pygmy chimpanzees are rich in individuality, and the personality of individuals probably exerts a strong influence on the character of social relationships between group members. Observations in recent years clearly show that the groups at Wamba differ greatly in social structure, and these group "personalities" are probably rooted in differences in the personalities of the members of each group. In the ongoing research, then, there is a great need to focus our attention on "personality."

For six or seven years following the commencement of our investigations in 1973, the forests of Wamba were a paradise for the pygmy chimpanzees and for us researchers. The pygmy chimpanzees, not bothered by the presence of humans, pursued a carefree life moving around the forest, and we followed them around without a care in the world. Beginning in about 1980, however, several threats developed. Wamba had gained fame as a forest where chimpanzees abound, and the poachers who began to enter took the lives of an unknown number of them. Worse, the central government of Zaire sometimes gave the order to capture and produce a chimpanzee youngster from the Congo forest region. Each time, the regional officials came to search the Wamba forest, because it was so much easier to capture a chimpanzee habituated to people than to catch one that was not.

The coffee trade, which was doing well, posed another threat. Because they had begun to prosper from the sale of their coffee beans, the villagers quickly began clearing the primary forest and planting more coffee. Quickly, the habitat of the Wamba pygmy chimpanzees began to be destroyed. On top of that, news came that a West German company had begun large-scale lumbering in the Congo forest, and that the Wamba forest was within the boundaries of their plans. It was crucial that we develop a plan for protecting Wamba's pygmy chimpanzees and, beginning in 1983, we persistently appealed. Finally, in 1987, Wamba Forest and the Luo River to the south, along with the neighboring Ilongo Forest, were set aside as a reserve for chimpanzees. There was now a legal basis for preventing new lumbering of primary forest and the capture of chimpanzees in this region.

But in 1988, during our researchers' temporary absence, regional officials once again entered Wamba Forest and captured chimpanzees.

That the formal establishment of a reserve was ineffective in checking the capture of chimpanzees, even by the regional government, came as a shock to me. The greatest disappointment was that some of the very people who were assigned the duty of protecting the chimpanzees, under the new law, had initiated their capture.

Through such depredations, the pygmy chimpanzees have been pushed to the edge of extinction. To protect them from extermination, we must make people aware of this precious ape. I hope that my book, along with Dr. Suehisa Kuroda's and the several contributions of Randall Susman, may play a role both in our understanding of the pygmy chimpanzee and in its conservation.

I am pleased and honored that this book is being published in the United States. Every year in Japan, the writings of many Western primatologists are translated into Japanese, but outside of scientific journals, there are relatively few opportunities for Japanese researchers to present their observations and thoughts to Western readers.

This is the first time that Evelyn Ono Vineberg has translated a Japanese work. In order to discuss the problems encountered in translating the book, she twice left her family to visit Japan. I want to express my heartfelt appreciation and admiration for her boundless enthusiasm and perseverance in bringing the translation and publication of this book to fruition. In the process, happily, I have gained a wonderful friend.

T. K.

Translator's Preface

Today many Japanese primatologists work diligently at their computers producing papers for publication in English. A few of these will be published, but the bulk of Japanese primatological literature will remain in Japanese and inaccessible to most Westerners.

This is an unfortunate situation. At a time when the forest homes of our closest nonhuman relatives are being destroyed, we ought to make every effort to pool our resources, in particular our knowledge of the animals we hope to conserve for future generations. Researchers and conservationists need access, through translation, to the important works of fieldworkers all over the world. It is unrealistic to expect non-English speakers to publish books simultaneously in English or to write only in English, thus excluding an audience of their own homelands.

By happenstance and the workings of politics, English has become the international language. But we English speakers must not forget that rich libraries of information are available in other languages. If science is to reveal truths about our world and about mankind, we must be able to assimilate the findings of scientists writing in other languages and to bring insights from many different cultures to bear on our understanding of humankind and our environment. The burden is on us to build up our own resources of translators and interpreters while continuing to encourage and facilitate the exchange of information in English.

With this sense of imbalance in the field of primatology, and fortified by personal interest and experience observing pygmy chimpanzees in captivity and in the wild, I decided to translate this book. It was November 1986, and I was introducing my friend Professor Suehisa Kuroda (Laboratory of Physical Anthropology, Kyoto University) to Dr. Don Lindburg (Research Behaviorist at the San Diego Zoo). Several years earlier, I had spent almost six months in Zaire with Dr. Kuroda, 3½

THE LAST APE

months of that at the Japanese field site at Wamba described in this book. Don told me about a special man, a close friend of the family of pygmy chimpanzees at the San Diego Zoo. Dr. Jim Diggins was willing to sponsor the translation of a book about wild pygmy chimpanzees for the benefit of our San Diego group and for all those people interested in knowing more about this fascinating ape. Don had a copy of Kuroda's book, which was written in 1982, but did not know about Takayoshi Kano's book, just published in 1986. Without realizing what I was getting myself into, I took up Kuroda's suggestion that Professor Kano's book be translated before his. Just that easily, I had naively agreed to translate two books about pygmy chimpanzees, the one you are about to read and Suehisa Kuroda's award-winning book *Pygmy Chimpanzee: The Unknown Ape* (*Pigumi chinpanji: Michi no ruijinen* [Tokyo: Chikuma-shobo, 1982]).

How does one begin to translate a book for the first time? Anyone who has tackled Japanese—the written language—knows what happens when you leave it for a while. The Chinese characters, the kanji, begin to drop out of your recall storage units. It is a case of use or lose, and after an interval of many years, I had apparently lost whatever I had.

In the initial stages, I invested scores of hours counting strokes and looking up kanji one by one in Nelson's character dictionary. Just when I despaired that I could never recall kanji again, the characters began to stick in my memory tracks. I must admit that I agreed with a few friends and family members who said that my endeavor was absurd, that it was bound to end in failure and loss of face. When I finally accepted these risks, I was able to overcome my paralyzing fears.

I requested sample translations of the first several pages of the book from three different people. One was a well-educated native Japanese who has lived in the United States for more than 15 years. The other two were Americans, a professional translator and a Japanese-language major, who had each spent more than four years living in Japan. Each of the three translated the initial pages of the book, and we all came up with different translations. The variation in transforming the Japanese expressions into English ranged from the quite literal to the quite liberal.

Given my limited funding, the slowness of the work (everyone estimated they had spent 1½ to 2 hours per page), and the nonuniformity in styles of translation, I realized that I had to tackle the book alone. My two early drafts were quite literal, but by the last draft (the fifth), I was trying to make the English as native-sounding as possible for the sake of the English-speaking reader while preserving the original intent of the author. Nevertheless, I take small comfort in R. A. Miller's (1986) words that "translation is always an open-ended process . . . and is not an operation that admits of final solutions."

Syntactically and semantically, Japanese is very different from English. There is no historical or "genetic" link between the two languages (Kindaichi, 1978). Though Japanese has often been described as a "vague" language, this seems to be a cultural tendency rather than a limitation of the language itself (Miller, 1986). Where Kano's ideas seemed unclear to me or to the Press's editor, I questioned the author directly. In most cases, I was able to clarify the points through correspondence or direct conversation with him. Sometimes, in reading over the English version, he added or deleted certain sentences or passages, the better to present the information, or clarified a point that in his own view was not clear in the original version. My editor at Stanford University Press, Bill Carver, encouraged modifications of this kind, although no attempt was made to update the basic contents of the book.

Dr. Kano's style of discourse throughout the Japanese version of the book was primarily passive. I consciously tried to reword passages to give the sentences an active voice in English. Also, Japanese abounds with expressions using double negatives to make assertions. For example, instead of saying something "rarely occurs," a sentence in Japanese would literally mean "it does not happen often." In most of these cases, I have changed the expression to a positive statement. I broke paragraph-long sentences in Japanese into shorter English sentences, according to my own sense of style and readability. In Japanese, relative pronouns are often omitted, since they are understood within the context of a sentence. In the English, I usually added these parts of speech. Finally, in Japanese, it is possible to refer to a previously mentioned idea by "sore wa," which translates to "that." In English, however, it is often necessary to repeat an idea to which "sore wa" refers. For a readable, comprehensive description of the Japanese language, I refer the reader to Kindaichi (1978).

In the course of translating this book, I searched the literature for information about translation in general and in particular. I discovered that problems of translation are current themes in ethnography (Agar, 1980; Carroll, 1988; van Maanen, 1988; Marcus and Fischer, 1986; Werner and Schoepfle, 1988), in Japanese language and literature studies (Miller, 1977, 1986; Petersen, 1979; Seward, 1983; Taylor, 1979), and in writing (Barzun, 1986; Barzun and Graff, 1977; Kline, 1978; Purves, 1988; Radice and Reynolds, 1987), although the process of translation remains little understood. I could not find an authoritative text to point to to say, for example, that I translated this book "using the Suzuki method."

Although Dr. Kano was a good correspondent, reading, correcting, and commenting on my English draft of his book, I felt very strongly that for many details regarding the book, we needed to interact face-to-face. In many translation projects, it is not possible to work directly with the author, but when one can, it seems logical to get as much of a sense of the author and his thinking as possible. Realizing that relationships between Japanese and Americans could not be established overnight, and that the author is a shy man, I invested my earnings from this project in two trips to Japan. In 1988, I was accommodated for a month by several Japanese primatologist friends and friends of friends, including Suehisa Kuroda, Tetsuro Matsuzawa, Keiko Nakatsuka, and Shun Sato. And in 1989, Professor Pamela Asquith and Naobi Okayasu and her parents graciously accommodated me for part of my one-month stay. These trips allowed me to develop a sense of the context in which this book was written.

Providing the lost context for the original Japanese version of this book could be the subject of another book. A foundation for understanding the traditional Japanese approach to primatology has already been laid, by Pamela Asquith (1981, 1986) and Jean Kitahara-Frisch (Frisch, 1959, 1963). Through the writings of Japanese primatologists, a reader can gain a sense of the similarities and differences between communities of primatologists. Some background ideas, however, may enable the reader to understand this work at more than one level. Many readers will examine this book primarily for the valuable information it contains about pygmy chimpanzees. This is the reason the book was written. At another level, it can be considered an example of Japanese research. At this level, context is all-important, because the way in which the book will be evaluated will depend to a large extent on the reader's ability to understand the author's point of view and on how well his ideas are substantiated. Therefore, the reader should bear in mind that first of all, this English version has not been significantly altered for a Western, English-speaking audience. That is, beyond the transformation of Japanese to English (which is substantial), the author and I have not tried to rewrite the book for a different audience. At the same time, although the reader may be struck by the use of anthropomorphic language, a characteristic of traditional Japanese primatological work (see Asquith, 1986), in certain descriptions of behavior, some effort was made to reduce its occurrence. For example, the translator had the option of using terms such as child or baby instead of juvenile, infant, or offspring when faced with terms such as "kodomo" or "akanbou." Japanese primatologists do not use a separate vocabulary for describing the offspring of nonhuman primates, but they do have terms, such as "mesu" and "osu," for male and female animals that differ from "onna" or "otoko" for female and male human beings. In the former case, I had a choice, depending on whether I wanted to make the author sound very anthropomorphic and whether that was his intent (it was not). In the latter case, I did not see a choice.

Perhaps here, in the context of choices, culture, and politics, I should note that the subject of this book, *Pan paniscus*, goes by two vernacular names, the pygmy chimpanzee and the bonobo. Japanese primatologists, however, have always used the term pygmy chimpanzee, and I have therefore used this common name throughout the text to reflect both the author's background and his choice of name.

The reader should keep in mind that we (Americans) take for granted the large number of books that are available to us in primatology. The more recent of these books, however, are not simultaneously produced in Japanese. Therefore, a Japanese author's ability to cite the latest literature in English may depend upon the facility with which he can digest English, his network of connections with Westerners, and the speed with which some English books are translated into Japanese. Just as we are ignorant of the latest works in Japanese primatology, so the Japanese are often not as well read in English as we might assume.

I remember sitting in the library at the Primate Research Institute one Saturday afternoon and hearing a professor in the next room speaking to himself in English, repeating the same phrase over and over with slightly different intonation. I did not interrupt him, but wondered if his efforts were like mine, reading aloud some section in Japanese repeatedly, trying to get the rhythm or sense of the expression by hearing it instead of just looking at the writing.

Communication via the medium of writing is totally different from speaking. Spoken words are ephemeral and spontaneous; written ones are permanent and measured. Communication, or the process of achieving shared meanings, is interactive in speech, but not in writing. A writer must have a sense of the reader without really being able to interact with him or her. He must make assumptions about how much knowledge the reader has and decide what needs to be stated explicitly and how much and what kind of context to create in his work (Martlew, 1983). This problem is universal, but in the case of a translated work, it is extremely important to have a sense of the readership for which the book was originally intended. In this case, as mentioned previously, I can only direct the interested reader to the work of others for a discussion of some of the major aspects of Japanese culture (e.g., Nakane, 1972; Lebra and Lebra, 1986; Matsumoto, 1988) and of the Japanese approach to primatology.

As I stare at the English words in this book, I doubt their meanings. There is no similarity between my English and Dr. Kano's Japanese writing, yet he and I agree, on the basis of our knowledge of each other's language, that what is said here is basically what was meant in Japanese.

Translation allows people to see a picture that they would not be able to see otherwise, but they see the picture through a filtered lens. The lens should not change the objects or subjects of the scene, but it will change the colors—some hues may fade out completely and others may be darkened or lightened. And so the "feeling" is likely to be different between the original and the translation. Accepting this important limitation, I was finally able to finish translating the book. As a novice translator I expect there will be errors and mistranslations, but I hope these will not lessen the value of this work in increasing our understanding of these fascinating apes.

In closing, I want to acknowledge the initial, unwavering, and wholehearted support of Dr. Jim Diggins for this project. His initial financial backing helped me to gain additional funds for the translation from the L. S. B. Leakey Foundation and from Stanford University Press. To Jim and the institutions who funded this endeavor, I express my deep appreciation.

Many other people were helpful in supporting my search for funding and in making my trips to Japan productive and less costly. I want to express my appreciation to all of them for making things easier along this long way: Pamela Asquith, David Hill, Mike Huffman, Hiroshi Ihobe, Ellen Ingmanson, Takayoshi Kano, Suehisa Kuroda, Don Lindburg, Tetsuro Matsuzawa, Steve McCallion, Jim Moore, Keiko Nakatsuka, Toshisada Nishida, Naobi Okayasu and her parents, Hideyuki Ohsawa, Dr. and Mrs. Shun Sato, Nancy Tedokon, Rie Tsuboi, Dan Whitney, Adrienne Zihlman, and many other faculty and students of Kyoto University's Laboratory for Human Evolution Studies, Laboratory of Physical Anthropology, African Area Studies Center, and Primate Research Institute.

I am extremely grateful for the helpful criticism and guidance of my two editors at Stanford University Press. Bettyann Kevles has been a tremendous support, both technically and spiritually, from the earliest stages of this work, and she has been a good friend from the time I met her in 1987. Bill Carver guided the publication of this book and stood by me when I doubted it would ever make it to press.

Most important in bringing this project to a successful completion have been all the members of my family, including my parents. My children, Thomas and Kerry, showed a maturity far exceeding their youthful ages, rarely complaining about a work schedule that often took me away from them, sometimes for long periods. My husband Mickey bore a larger burden of responsibilities at home to give me additional time to work. Not only did he graciously play the role of house-husband during my absences, he carefully read and critiqued much of the manuscript, giving me additional valuable feedback. My parents as always supported me by giving of themselves, caring for my children, and enabling me to travel without being loaded down by worry. I owe my greatest debt of gratitude to this special family, all of whom accepted and honored the importance of this work to me.

To Kakowet, Linda, Laura, and Lorel, the original family of pygmy chimpanzees at the San Diego Zoo who captivated and motivated me to enter the field of primatology, I also owe deep thanks. I hope that having this book available in English will in some way improve the chances of their relatives' survival in the wild as well as in captivity.

E. O. V.

Contents

ONE

Why Study Chimpanzees?

I

TWO

Distribution

39

Contents

Figures & Maps

Figures

Maps

Seventy-eight unnumbered photos illuminate the text

Tables

THE LAST APE

*Pygmy Chimpanzee Behavior
and Ecology*

CHAPTER ONE

Why Study Chimpanzees?

Parallels with Ourselves

Human beings, like most other primates, are social animals. An individual cut off from others has difficulty surviving. Consequently, the regulation of relationships between individuals and between groups is of the utmost importance to human existence. To that end we created our various social systems, rules, customs, laws, and religions.

We cannot, however, use these social institutions the way we would use machines. It is impossible to find any human interaction in which emotions do not carry some weight, even during relaxed, polite conversation. Hierarchical relations, spheres of influence, personal attachments, desires, hostilities, jealousies, racial consciousness, and so on play roles in any kind of interchange. It is no exaggeration to say that all of these factors support and influence human social behavior, from small friendly associations between individuals to relations and treaties between nations.

All primate species except the orangutan and some prosimians live in social groups. The emotions and behaviors of primates have evolved through natural selection to support the social group. Likewise, human emotions and behaviors are products of evolution supported within social groups. Consequently, many rules, laws, and systems in human societies are meant to justify our animal emotions and behaviors—and to prevent us from abandoning them rather than to restrain us from revealing them.

Incest avoidance is a good illustration of this. There are differences in the extent to which consanguineous marriage is permitted, depending on race and tribe. In nearly all human societies, however, incest is challenged. For example, first-cousin marriages are socially permitted in Japan and India. Those in Japan amount to 4–6% of all marriages, and the proportion in India reaches 10%. Nevertheless, marriage between

The subjects of this study, at ease in the trees.

closer relatives is prohibited in both countries. In Europe and America, the incidence of first-cousin marriages is much smaller than in Japan and India, and African Bantu tribal society is still more strict, with marriage between traceable blood relatives forbidden by custom. In these African societies, anyone who marries a close relative will be reproached and branded as a "beast" (although animals would no doubt consider that characterization an affront).

Psychological and social mechanisms for avoiding incest exist in all species of primates. According to Junichiro Itani (1972), primate society has evolved along an axis of "incest avoidance." For example, males, females, or members of *both* sexes leave their natal group upon reaching sexual maturity. As a result, mother-son and father-daughter matings are prevented. Far from being a unique human-produced, advanced behavior, incest avoidance is a behavior that arose in the course of primate evolution.

In human societies, incest avoidance is either amplified or reduced, depending on the social demands of particular groups. There have been instances when incest was practiced frequently or was even encouraged. For example, in the ancient Egyptian pharaoh's household, brother-sister marriages were permitted as regular marriages. In Japan, too, consanguineous marriage was common among the ancient royal and aristocratic classes. In both instances, the promotion of incest occurred only in royal or aristocratic families where family lineages were considered sacred. Those families, which were socially isolated groups, were deliberately estranged from their fundamental human nature. Incest avoidance is indeed an inheritance from primates that existed long before humans.

Much of human behavior probably has a similar biological foundation. In anthropology and primatology, considerable effort has gone into exposing the roots of these behaviors and toward elucidating the process of evolution. At the same time, these efforts may help us understand ourselves.

Two traditional methods of research are used in these endeavors. One is to study the paleontological record. This approach is called the direct method, because it investigates the bones and traces that reflect the life style of animals that lived in the past. The goal is to reveal the morphology and behavior of past species from reconstructed, ancient environments by analyzing remains such as fossils, footprints, tools, food, pollen, and so forth.

Unfortunately, the conditions under which such remains are left are limited. For instance, bones fossilize less readily in the accumulated soils of forests, where primates occur, than in volcanic ash. As a result, primate fossils are few in number. Furthermore, fossils deteriorate with time, becoming more fragmentary and imperfect as they age. Yet, until a time machine is invented, the real life and behavior of extinct species will be impossible to view by any other means. Instead, we will have to rely on supposition based on the meager direct evidence left to us.

The other research method, which may be called "indirect," is to compare the structure and behavior of extant animals to those of *Homo sapiens* in an effort to give substance to the fossil evidence. Two ways to do this are by experimental laboratory research on various primates and by field research on the animals in their natural state. In laboratory experiments set up with specific research aims, a scientist has the advantage of being able to control in various ways the living conditions of an animal and the kinds, timing, and quantity of stimulus-induced behaviors. If planned carefully, such experiments can yield results in a comparatively short time. On the other hand, laboratory experiments have little power to inform us of the kinds of behavior that animals display in the so-called natural state.

Field research is exactly the opposite. Observation conditions are generally much poorer than in a laboratory. Most wild animals will either run away or create a commotion if they see a human. Certainly they will not act naturally at first, although there are some exceptions, such as the Indian rhesus monkey and the hanuman langur, which have a history of

The feeding site, where pygmy chimpanzees are provisioned with sugarcane, seen from the observation hut. As many as 20 to 25 chimpanzees are often in attendance.

At the feeding site, taking sugarcane.

harmonious relations with humans. There are ways of observing from complete concealment, but such observations are severely limited, no matter how many years are devoted to them, because the animal cannot be followed throughout its environment.

The study of wild primates begins only after the subjects grow ac-

] 4 [

The observation hut, built atop an old termite mound, with the feeding site in the foreground.

customed to the investigator's presence. This period of gradual "habituation" at the beginning of field research is so long (often two to three years) and painful that the investigator can lose patience. Even after habituation is achieved, experiments concerning the intricate, intertwined, underlying ecological conditions can be extremely limited. Still, only observations of primates in the natural state will tell us how they survive independently or as a group in a given environment. Thus field investigations, although difficult, are essential to research on the evolution of the behavior and society of primates.

"Provisioning" is a method that has been used, mainly by Japanese researchers, to facilitate habituation. Many reject this method because it encourages deviations from natural behavior, but because provisioning can quickly eliminate an animal's apprehension of humans, it is an effective method of investigation if used cautiously.

Researchers interested in the origins and evolution of human behavior may seek to infer our own history from the surviving traits of other species. Inevitably such extrapolations are taken as standards, but unrestricted extrapolation holds the danger of absurd errors. The choice of a subject group as closely related to humans as possible is advisable because close relatives more than distant ones tend to retain qualities resembling those of our common ancestor. Because humans belong to the taxonomic

Table 1. Classification of recent primates
(Napier and Napier, 1967; Zihlman, 1982)

Order Primates
  Suborder Prosimii (lower primates)
    Superfamily Tupaioidea
      Family Tupaidae (e.g., tree shrew)
    Superfamily Tarsioidea
      Family Tarsiidae (e.g., tarsier)
    Superfamily Lorisoidea
      Family Lorisiidae (e.g., galago, loris)
    Superfamily Lemuroidea
      Family Lemuridae (e.g., lemur)
      Family Indriidae (e.g., indri, sifaka)
      Family Daubentoniidae (e.g., aye-aye)
  Suborder Anthropoidea (higher primates)
    Infraorder Platyrrhini
      Superfamily Ceboidea (New World monkeys)
        Family Cebidae (e.g., squirrel monkey, capuchin)
        Family Callitrichidae (e.g., marmoset)
    Infraorder Catarrhini
      Superfamily Cercopithecoidea (Old World monkeys)
        Family Cercopithecidae (e.g., macaque, guenon)
        Family Colobidae (e.g., langur, colobus monkey)
      Superfamily Hominoidea (apes and humans)
        Family Hominidae (e.g., modern humans)
        Family Pongidae (e.g., chimpanzee, gorilla)
        Family Hylobatidae (e.g., gibbon)

order of Primates (Table 1), research on primates extrapolates to humans better than does research on non-primate groups.

In addition to taxonomic affinities, we must also seriously consider similarities in environment. For example, one clear method for showing common behavioral traits is to select animals that live in conditions resembling those to which our ancestors became adapted. This method, however, is dangerously prone to environmental determinism if we ignore such matters as lineage. If the lineages differ but the environments are the same, the adaptations likely evolved in different ways. Therefore, even if importance is attached to common environment, a subject should be chosen from as close a class of relative as possible.

The ancestral model of primates is an ancient prosimian type that appeared as much as 70 million years ago (mya). Its teeth and skull resembled those of a primate, but its overall external appearance was very different. The nose jutted out, the eyes were small, and the fingers were hooklike. It was probably a forest dweller that occupied a niche close to the ground, and it may have been nocturnal.

Between 60 and 50 mya, the prosimians increased in diversity and greatly extended their geographic range. During this era, the principal thrust of the evolutionary direction of primates was determined. In some prosimians, anatomically important changes took place. The claws were replaced by flattened nails, the fingertips became well-padded tactile organs, and creases in the palms made grasping objects easy. In some primates, protrusion of the nose (nasal prognathism) was reduced, and the face became flattened. As the eyes began to face forward the field of vision narrowed, but this change enabled primates to view their world in three dimensions. These were epoch-making adaptations to life in the trees. The accurate measurement of distance using stereoscopic vision enabled primates to move rapidly and safely through the space where trees and branches intersect. Together with the revolution in the mode of walking, these adaptations in the hands and vision of primates opened the road to human tool-making (that is, using the hands to make tools).

In this Golden Age, the prosimians were not confined to Asia and Africa, as they are now, but lived also in Europe and America. Somehow they became extinct in Europe and America and, today, glimpses of their showy past endure only in Madagascar. In Asia and parts of Africa, they were shoved into a corner by the emergence of true monkeys (suborder Anthropoidea). There, the prosimians barely survived by foraging at night when the monkeys were inactive.

The diversification of anthropoids (monkeys, apes, humans) occurred between 50 and 30 mya. This is called the second-order radiation of the Primates, during which they expanded their habitat to encompass the upper story of the forest (*Colobus*, etc.) as well as the terrestrial environment (baboons, patas monkeys, etc.).

There are two groups of anthropoid monkeys, the New World monkeys and the Old World monkeys. The former are found in South America; the latter underwent a separate radiation in Asia and Africa. Both groups of anthropoids developed advanced stereoscopic vision and acquired a sense of color. They displayed increased manual dexterity, and their thumb became opposable. They could even move each finger separately.

Apes were the last group to radiate from the prototype of anthropoid monkeys. Humankind emerged from the apes. An ancestral form of ape has been discovered from a layer in Fayum, Egypt, that dates from approximately 27 mya, but the appearance of the great apes' distinctive

] 7 [

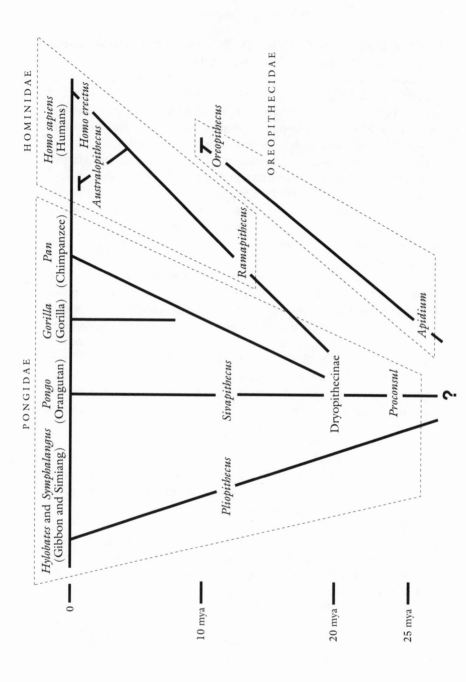

Figure 1. Classical hominoid family tree (after Simpson, 1963; Schultz, 1966).

modes of travel, "brachiation" and "knuckle-walking," is thought to have occurred much later, at about 20 mya (Fobes and King, 1982).

Brachiation is a form of locomotion in which an animal hangs by its arms from a branch and moves by swinging or alternating the arms. A relatively heavy animal can move and feed in the trees because brachiation redistributes body weight. The complete lack of a tail in apes is related to the change to brachiation: monkeys use the tail for balance when leaping and running, whereas brachiation does not require one. The shoulder, elbow, and wrist joints of apes are flexible and can rotate to a high degree. From a hanging position, various courses of movement are possible.

Knuckle-walking is another distinctive mode of travel. Among the apes, the gorilla and chimpanzee often walk on the ground by thrusting against the surface with the back of the middle part of their slightly bent knuckles. Formerly, knucklewalking was thought to arise from brachiation. Now, however, many people agree with Tuttle (1969), who thinks it evolved for terrestrial travel over long distances in the forest.

The African chimpanzee and gorilla and the Asian gibbon and orang-utan are the only surviving apes. Moreover, all of them are on the brink of extinction. Although the apes were the last to radiate, their reign of influence has rapidly declined, and only scatterings of these primates remain. They played an extremely important role in the biosphere, however, because we humans arose from an ape stock.

Although our ape ancestry has long been recognized, there is currently much debate about when the split between ape and human lineages occurred. Apes and humans are combined in the superfamily Hominoidea, with humans (*Homo sapiens*) placed in a separate family, the Hominidae. Within the Hominidae, there was an extinct group called *Ramapithecus* that until recently had been thought of as the oldest form. *Ramapithecus* and *Dryopithecus* (thought to be the ancestor of chimpanzees and gorillas) may have diverged in the Miocene, and certainly by the Pleistocene, the march down the road leading to humankind had begun. Thus, we thought that the divergence of gorilla, chimpanzee, and human may have occurred 20 mya (Fig. 1).

In the latter half of the 1960's, however, scientists who applied the research methods of biochemistry and molecular genetics raised objections to this view of the timing of divergence. First, using a special technique for determining immunological distances between species through the comparative analysis of the protein albumin, V. M. Sarich and A. C. Wilson (1967) calculated the time of divergence between groups of primates. Then, using the paleontologically accepted time of 30 mya for the split between Old World monkeys and New World monkeys, they obtained results that indicated 5 mya for the divergence of the African apes (chimpanzee and gorilla) and man. From the immunological distance measured using transferrin (a type of blood serum protein), they calculated practically the same time of divergence for the African apes, 4.6 mya. Second, M. C. King and A. C. Wilson (1975) discovered that humans and chimpanzees, if treated as common mammals, differ only to the degree of sibling species, on the basis of electrophoresis and the results of a comparative analysis of their proteins. Third, DNA analysis (DNA hybridization method) has made it clear that humans and chimpanzees are very closely related. According to D. Kohne and associates, DNA variation did not exceed 2.4%, and according to B. Hoyer's results from an identical experiment, the variation was no more than 1% (Takenaka, 1983). According to Takenaka, these experiments show that the divergence of humans and chimpanzees occurred 4.3 mya. Ape and Old World monkey divergence is assumed to have been 35 mya.

The current, consensus view of molecular biologists is that the divergence time of humans and chimpanzees was 4.5–5 mya (Sarich, 1984). By contrast, paleoanthropologists had formerly stated that this divergence occurred as much as 20 mya. What factors created a difference of this magnitude? The extrapolation of data from the evolutionary systematics of molecular biology has some limits. Among them, the findings simply give an indication of the divergence times of the two types. They are powerless to show the physical appearance of the common ancestor of the two lineages. For that purpose, fossils are necessary, and the time scale of the molecular clock must be calibrated, like a "stopwatch." For example, V. M. Sarich and his associates regard 30 mya as the divergence time of Old World monkeys and New World monkeys, and they calculate the divergence time of each kind of primate on the basis of that. But what if the initial established time is mistaken?

Moreover, there is the actual fossil that we call *Ramapithecus*, which has been regarded as the direct ancestor of mankind. If its period of survivorship is 8–12 mya (according to a combination of the potassium-argon method and others, we are able to measure the period with fair accuracy), there is no reason to deny the importance of *Ramapithecus* as a likely ancestor. In the 1980's, however, well-preserved fossils of *Ramapithecus* were discovered in Pakistan and elsewhere. Discussion revolved around the evidence (Andrews and Cronin, 1982) that *Ramapithecus* belonged in the same taxon with *Sivapithecus*, linking it to the orangutan and not to man. In paleoanthropology, *Ramapithecus* had lost its importance as a candidate for the oldest human ancestor. When *Ramapithecus* is removed from the hominid group, the oldest remaining fossil is *Australopithecus*, and among the australopithecines, even the oldest is not more than about 3 million years old. Thus, the paleontological basis for denying the results of the molecular clock vanish with *Ramapithecus*.

In 1982, Hidemi Ishida discovered a fossil in Kenya that belongs to the Hominoidea and dates from 7 to 9 mya. This fossil has not yet been named in proper form, but has been temporarily named Sanburu Hominoid. Japanese researchers hoped that it might be one of the oldest examples of humankind. Between *Australopithecus* and *Ramapithecus*, there is a long blank record of hominoid fossils, and Ishida's discovery merits great attention as something to fill that gap. Preliminary results, however, indicate that the Sanburu Hominoid does not belong to the Hominidae. Instead, Ishida thinks that it may be close to the common ancestor of man and the gorilla-chimpanzee lineage (Ishida, pers. comm.). In any case, it does not contradict the evidence of the molecular clock.

The australopithecines, which are man's oldest ancestors, are divided into three main lineages, *Australopithecus afarensis*, *A. africanus*, and *A. robustus*. They were all savanna residents. The lineage of *A. afarensis* made progress in tool manufacture and use, and gradually proceeded to *Homo sapiens* by way of *H. habilis* and *H. erectus*. Thus, I consider *A. afarensis* to be the lineal descendant of the root stock leading to humans. Meanwhile, the other two australopithecine lineages, which showed no progress in tool manufacture and use, and had specialized food habits, failed to adapt to a severe savanna environment and became extinct (Fig. 2).

Recently, an interesting editorial in *New Scientist* magazine objected to this common opinion. It argues that *A. africanus* and *A. robustus* did not simply vanish; rather, they returned to the forest from the savanna and became chimpanzees and gorillas (Gribbin and Cherfas, 1981). For two widely distributed lineages of *Australopithecus* suddenly to have gone extinct is unnatural. The fossil forms of Africa's ancient apes (dryopithecines) suddenly disappear more than 10 mya. It is strange that suddenly apes (chimpanzees and gorillas) show up again after the extinction of the two species of *Australopithecus*. Australopithecines were bipeds, but it is not impossible that they shifted from bipedalism to chimpanzee and gorilla brachiation and quadrupedalism. Certainly, this is an extraordinary theory, because it asserts that after starting down the road toward man, humanlike forms that returned to the forest reversed course and became the chimpanzee and the gorilla.

Nevertheless, a similar theory, based on work in molecular anthro-

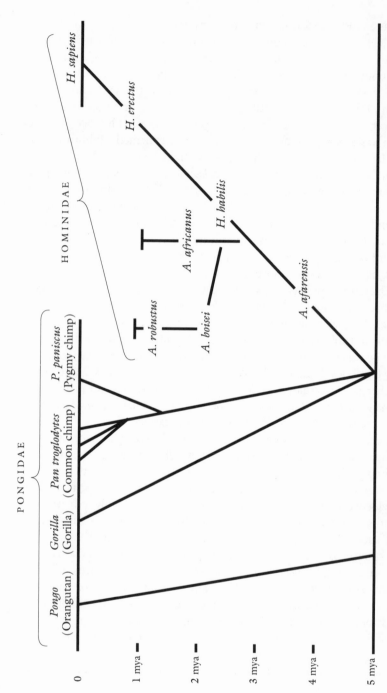

Figure 2. A phylogeny of hominids and great apes based on paleontological evidence and results of DNA analysis. (*H: Homo, A: Australopithecus.* For hominids: Tobias, 1965; Johanson and White, 1979; White, 1983. For gorilla and chimpanzee: Sarich, 1984. For orangutan: Sibley and Ahlquist, 1984.)

pology, was published recently. Masami Hasegawa (1984) asserted that, according to mitochondrial DNA analysis, African apes come *after A. afarensis*. He says that the chimpanzee and gorilla as well as the human and *Australopithecus* should be lumped into the same subfamily Hominidae (Fig. 3), and the difference between humans and African apes is only at the genus level. Parties that had been thought to be distant relatives now had come to be identified as half-brothers. There is still, however, scarcely any reason for this view to be accepted. Nevertheless, the perceived distance between man and chimpanzee has been narrowing swiftly ever since evidence from molecular biology began to accumulate. This is encouraging to researchers who study chimpanzee behavior. Probably more than any other idea, this new theory—that for a period the behavior of chimpanzees and the behavior of humans shared the same genetic foundation—has greatly reduced the risk of erroneous extrapolation of information from one to the other.

Several experiments in behavioral research on non-primate animals have elucidated the biological foundations of human behavior. Among these, Konrad Lorenz's (1966) classic study is the most famous. On the basis of behavioral research on fishes and birds, he concluded that aggression is the foundation, as well as the starting point, of every possible social relationship. According to Lorenz, by preventing the exhaustion of limited resources, aggression not only permits the coexistence of kindred individuals but is essential for true love and group cohesion. Certainly, this is true for human beings too.

Robert Ardrey's (1961, 1966) view is similar, but he regards territoriality as more important than aggressive instincts. Aggression is thought to be an expression of territorial instincts. Human beings have territorial instincts that developed so that humans would fight to defend their territory. Group cohesion and friendship were born from these instincts. Territoriality in humans may have evolved at the time when *Australopithecus* moved to the savanna and began hunting there. Territoriality may be an instinct we have had for millions of years and may not be able to extinguish.

Ardrey (1970) also says that a system of rank has the greatest importance historically. Almost all primates that form social groups, and many other mammals, have a rank system based on dominant-subordinate relationships. Certainly human beings are not exceptional in this regard. Ardrey emphasizes that a class society built on a rank system is a stable one, in which everyone can coexist.

Masao Kawai (1979) focused on the phenomenon of conspecific kill-

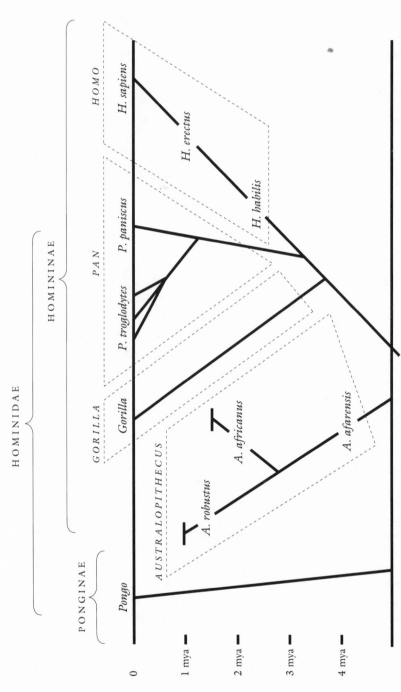

Figure 3. New version of hominoid systematics. (Hasegawa, 1984; for chimpanzees, Sarich, 1984. The means for the divergence times estimated by Hasegawa, 1984, were used. The divergence of chimpanzees from humans was 1.90–4.1 mya, and the gorilla family split off 2.8–5.9 mya.)

ing, which is widely seen in primates. Primates make their living in forest trees and were the first mammals to achieve great success there. In the trees, food is abundant, and there are few enemies or competitors. Primates radiated as if rushing into an empty field. They had to confront, however, the enemy within, called "over-population," which is an accompanying enemy of radiations. In the absence of natural enemies, they had to control their own numbers, by themselves, by conspecific killing. War occurs repeatedly throughout human history and may have an evolutionary basis in conspecific killing. If future populations continue to increase, war may be unavoidable.

The theories of Lorenz, Ardrey, and Kawai look upon aggression as the biological basis for human behavior. They have in common the view that coexistence and survival are designed according to man's interpersonal aggressive nature.

The family, consisting of a male-female pair together with their children, constitutes the smallest human social unit. It appears to be common to all people no matter what their race or tribe; it is a basic feature of human beings. In nonhuman primates, the males barely participate in the care of offspring, and the female, who is usually slighter in build, plays the inequitable role of long-term caretaker. In contrast, during human evolutionary history, the male began to visit the mother and child and to participate in child-rearing in the form of economic support. This was an epoch-making event.

Also in the course of evolution, the human brain continued to enlarge. Meanwhile, because there was a limit to the enlargement of the female pelvic canal, it became necessary for the female to deliver the baby in an immature state in order to give birth safely. This trend of decreasing maturity at birth was spurred on by the increasing size of the brain. As the trend progressed, childcare qualitatively increased in difficulty, the period of dependency continued to lengthen, and the burden on the mother increased. If the males had not joined the mother-infant unit, that is, if they had not formed a family, humans probably could not have survived in the harsh savanna environment.

The structure of primate society is not simply fixed according to the species, but resembles that of other closely related species. The social group changes through time and goes through transitions. When deviations occur, however, various controls are at work to return the group to the original social structure. These patterns of "social homeostasis" also resemble each other among closely related taxa. The indication is that heredity plays an important role in structuring primate society although

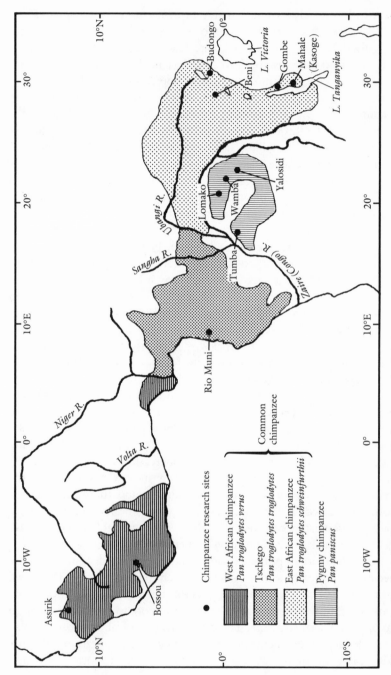

Map 1. Distribution of chimpanzees (Kortlandt, 1983, and others).

there are considerable variations derived from environmental differences. In other words, social structure is a product of evolution the same as features of anatomy, physiology, or behavior.

The family should also have a social structure that has evolved. At present, however, no element suggesting the origin of the human family has been found in any nonhuman primate society, including that of the apes. Nonhuman primate society and the human family appear to be completely separated, and the distance between them is as great as that between quadrupedalism and bipedalism. The clue to filling this gap is undoubtedly hidden in the social structure of apes, especially chimpanzees.

The Discovery of Chimpanzees

Chimpanzees (*Pan*) comprise two species, *Pan troglodytes* and *P. paniscus*. The former is further divided into three subspecies (*P. t. verus*, *P. t. troglodytes*, and *P. t. schweinfurthii*) that have an extensive distribution reaching from the West African coast, through the forest belt of the north bank region of the Zaire (Congo) River, to the periphery of East Africa's Lake Tanganyika (Map 1). The three subspecies together are called, in general terms, "common" chimpanzee. Chimpanzees in domestic zoos, circuses, and so on are, almost without exception, this common type. The latter species of chimpanzee is limited to a range on the south side of the Zaire River and is called by various names, including bonobo and pygmy or dwarf chimpanzee. In this book, I shall use the most general term, pygmy chimpanzee.

The pygmy chimpanzee is a rare species. One can quickly count the places where they are bred in captivity: Zaire's Mobutu Zoo in Kinshasa, Belgium's Antwerp Zoo (Belgium once ruled Zaire), Germany's Frankfurt Zoo, and, in the USA, the San Diego Zoo, Yerkes Regional Primate Research Center, Milwaukee Zoo, and Cincinnati Zoo. As far as I know, only one juvenile pygmy chimpanzee is being raised in Japan, at the Rakutenchi Zoo in Oita, Kyushu.

Since ancient times, Europeans seem to have known of chimpanzees. The first record of something resembling a large-sized African ape goes back to 500 B.C. By one account, a man named Hanno from Carthage traveled south along the west coast of Africa and brought home skins of "wild men" called "gorillae" by the local inhabitants. We are unable to judge whether they were human beings, gorillas, or chimpanzees, but R. M. and A. W. Yerkes (1929) suppose that the skins were from either a chimpanzee or an extinct ape.

Although African apes are not mentioned in the literary works of Aristotle, Plato, and Galen, H. Johnston (1922, cited in Yerkes and Yerkes, 1929) provides evidence that the Greeks who settled in Mediterranean Egypt had observed chimpanzees. According to Johnston, several art works depicting what look like chimpanzees are in the Tanagura Collection (500–600 B.C.) of the British Museum. Reported to be in the collection is "an ape riding an ox," with the ape having features of the chimpanzees of East Africa (*Pan troglodytes schweinfurthii*).

After the sixteenth and seventeenth centuries, clear accounts appear of animals thought to be African apes. For example, there is the account of an Englishman named A. Battell who was captured by Portuguese in South America. Battell was sent directly to the prosecutor in São Paulo, Angola, the Portuguese colony on the west coast of Africa. During the period of his captivity, Battell heard stories from local people saying that two kinds of "monsters" inhabited the forest. His report was published in 1625 and a summary of the section about the monsters is given below (after Yerkes and Yerkes, 1929: 42–43).

> The Woods are so covered with Baboones, Monkies, Apes, and Parrots, that it will feare any man to trauaile in them alone. Here are also two kinds of Monsters, which are common in these Woods, and very dangerous.
> The greatest of these two Monsters is called, *Pongo*, in their Language: and the lesser is called, *Engeco*. This *Pongo* is in all proportion like a man, but that he is more like a Giant in stature, than a man: for he is very tall, and hath a mans face, hollow eyed, with long haire vpon his browes. His face and eares are without haire, and his hands also. His bodie is full of haire, but not very thicke, and it is of a dunnish colour. He differeth not from a man, but in his legs, for they haue no calfe. Hee goeth alwaies vpon his legs, and carrieth his hands clasped on the nape of his necke, when he goeth vpon the ground. They sleepe in the trees, and build shelters from the raine. They feed vpon Fruit that they find in the Woods, and vpon Nuts, for they eate no kind of flesh. They cannot speake, and have no vnderstanding more than a beast. The People of the Countrie, when they trauaile in the Woods, make fires where they sleepe in the night; and in the morning, when they are gone, the *Pongoes* will come and sit about the fire, till it goeth out: for they have no vnderstanding to lay the wood together. They goe many together, and kill many *Negroes* that trauaile in the Woods. Many times they fall vpon the Elephants, which come to feed where they be, and so beate them with their clubbed fists, and pieces of wood, that they will runne roaring away from them. Those *Pongoes* are neuer taken aliue, because they are so strong, that ten men cannot hold one of them: but yet they take many of their young ones with poisoned Arrowes. The young *Pongo* hangeth on his mothers bellie, with his hands fast clasped about her: so that, when the

Countrie people kill any of the femals, they take the young one, which hangeth fast vpon his mother. When they die among themselues, they couer the dead with great heapes of boughs and wood, which is commonly found in the Forrests.

The Pongo and Engeco in Battell's account are thought to be the gorilla and the chimpanzee, respectively. Although some people regret that his narrative about Pongo and Engeco has been forgotten, the account does confuse the two types of apes. For example, fruit and walnuts are chimpanzee food, not gorilla food. Also, nests are described as being made in the trees, whereas gorillas live on the ground, and in many instances make nests close to the ground. The story concerning chimpanzees and a blazing fire is a folktale that is widely told in the African forest belt. The people at my research site in Wamba tell another legend, this one about chimpanzees that walk on two legs with both hands placed around the neck at a place where they are not seen by people. It seems that Battell's account about Engeco was not forgotten; he had merely recited all the stories about Pongo and Engeco that he had heard from African people.

Battell's account does contain some truth about African apes. It seems, however, that zoologists of Battell's time completely disregarded it as a tall tale fabricated by a traveler. N. Tulp, a Dutch doctor, wrote the first account about chimpanzees that was "scientifically" witnessed. In 1641, he reported on an ape brought from Angola, calling it "Orangutang." Those who have seen the illustration and written description of this ape and know the producing area believe that the ape was a chimpanzee.

In 1699, an English doctor, E. Tyson, brought in a live ape, also from Angola. Soon after the young ape died, Tyson's friend W. Cooper was asked to help dissect it. "Orangutan and Pygmy" are the names they entered. According to Tyson, his Pygmy resembled a human more than a monkey in 47 characteristics, resembled a monkey more than a human in 34 characteristics, and was an animal halfway between a monkey and a human. Tyson expressed this bold and correct opinion, based on dissection, approximately 200 years before the publication of Darwin's *The Descent of Man* (1871). Judging from the drawings left by Tyson, it is clear that this ape was a common chimpanzee (Fig. 4).

Throughout the eighteenth century, the chimpanzee continued to be confused with the orangutan. During that time, ape specimens brought to Europe were of chimpanzees only. Information about the other apes, however, namely the gorilla, orangutan, and gibbon, was beginning to

Figure 4. Old records of chimpanzees. *Left:* drawing by Tulp, clearly a chimpanzee (from Tyson, 1699). *Right:* Tulp called this an orangutan, but it is a chimpanzee; there is webbing between the second and third toes (from Tulp, 1641).

be collected from travelers' accounts. These four types of apes were clearly distinguished after the mid-nineteenth century.

According to J. F. Blumenbach, the type specimen of the common chimpanzee was described in 1779, but not until early in the twentieth century did the distribution and rough location of chimpanzees become clear. Many specimens from various places were collected, and classification also advanced. In a review written in 1913, D. G. Elliot suggested that the genus *Pan* comprised ten main types, and added that there were undecided types.

The pygmy chimpanzee, however, was not yet included in this group. The specimen collector H. Johnston (1922), with his abundant experiences in Africa, complemented Elliot's information on distribution with numbers and sizes, but also said, "Certainly the gorilla and apparently the chimpanzee are not found west or south of the main stream of the Congo."

Reports of the existence of chimpanzees south of the Zaire River gradually began in the 1920's. As far as I know, however, the first written reports were published, separately, in 1928 by H. Schouteden and E. Schwarz. The next year, Ernst Schwarz described a new subspecies, *Pan satyrus paniscus*, using as a type specimen a female collected by J. Jesquielle from the zone of Befale, in the equatorial region of the Belgian Congo (present-day Zaire). This was the pygmy chimpanzee, and the time was exactly 150 years after the first description of the common chimpanzee.

Before this time, some pygmy chimpanzee specimens were undoubtedly brought into Europe and America, being mixed with specimens of the common chimpanzee. Vernon Reynolds (1967) and others, after examining Tulp's illustration and description, wondered if it was a pygmy chimpanzee. Reynolds gives three reasons for this view. First, the collection site was Angola; that is, it was south of the Zaire River. Second, a partial fusion or a membrane between the skin of the second and third digits of the right foot was apparently illustrated, and interdigital webbing is considered one of the distinctive features of the pygmy chimpanzee. Third, the specimen in Tulp's account is too small to be an adult female common chimpanzee, and its height is about that of a three-year-old juvenile common chimpanzee. A. Gijzen (1974) emphasizes other points of similarity with the pygmy chimpanzee in addition to the webbed toes, including its facial features, the way the facial hair grows, and the shape of the face. But she also comments that, except for the head, the body of the specimen is too stout and massive for a pygmy chimpanzee and looks more like that of a common chimpanzee.

The study of skins and photographs has confirmed (Bree, 1963) that a chimpanzee named Mafuka, raised at the Amsterdam Zoo from 1911 to 1916, was a pygmy chimpanzee. Mafuka apparently stood out among common chimpanzees for his behaviors and features. According to Gijzen (1974), A. F. J. Portielje (1916) described this chimpanzee in a guidebook of the Amsterdam Zoo as "an aberrant form of the chimpanzee, probably a new subspecies."

Judging from the photograph and descriptions by Robert M. Yerkes, a prominent scholar of the great apes, a chimpanzee named Prince Chim, raised at Yale University from 1923 to 1924, was undoubtedly a pygmy chimpanzee (Coolidge, 1933). Yerkes had presumed from its capturing site, Lubutsu (a town northwest of Lake Kivu), that Prince Chim belonged to *P. t. schweinfurthii* (*Pan schweinfurthii marugensis*). He thought that Prince Chim was probably five to six years old because the upper and lower first molars were permanent teeth. Yerkes also raised a female West African-born chimpanzee, named Panzee, with Prince Chim and did comparative observations. The two chimpanzees were conspicuously different in various external features and behaviors. According to Yerkes and Learned (1925), Prince Chim was "sanguine, venturesome, trustful, friendly, and energetic," whereas Panzee was "distrustful, retiring, and lethargic." In Yerkes's other book, *Almost Human* (1925), he again praised the pygmy chimpanzee: "Doubtless there are geniuses even among the anthropoid apes. Prince Chim seems to have been an intellectual genius."

Delforge, a Belgian who lived at Elizabetha, a town on the north shore of the upper Zaire River, came into possession of one juvenile chimpanzee from the south side and one from the north side of the river, and raised them for some time; his findings (Schouteden, 1931) were similar to those of Yerkes. According to Schouteden (1931), Delforge observed that the chimpanzee who came from the south side (that is, the pygmy one) had a smaller body than the one born on the north side (the common one). Nevertheless, the one from the south side understood more than the other and was a very agile and shrewd "fellow" who was treated exactly like a "servant-boy."

Both Yerkes (1925) and Delforge (Schouteden, 1931) point out differences between the voices of the two chimpanzee types. The vocalization of a chimpanzee is difficult to characterize in writing, but Delforge reports that the *troglodytes* (common type) vocalization "hu-hu-hu," contrasts with the *paniscus* (pygmy type) cry "hi-hi-hi." Put in other terms, the voice of the pygmy chimpanzee is a high, pointed, "castrated" sound that contrasts with the low, deep, resonant sound of the common chimpanzee. Toshisada Nishida, who heard the vocalization of a wild pygmy chimpanzee for the first time while visiting our research site at Wamba, described his impressions: "It was clearly a bird. However, when I listened harder, it was unmistakably the voice of a chimpanzee." This is an apt description.

The Place of Pygmy Chimpanzees

According to Ernst Schwarz (1929), the distinctive features of the pygmy chimpanzee type specimen are its small physical size, an enlarged frontal bone and rounded occiput, a small supraorbital torus, and lack of a sagittal crest (a bony protuberance running along the median line, in a fore-aft direction, on top of the skull). Schwarz (1928, 1934) also pointed out that all of these features of the pygmy chimpanzee were "infantile." He thought, however, that the differences from the common chimpanzee were limited to the subspecies level, and consequently, he gave the pygmy chimpanzee the name *Pan satyrus paniscus*.

Several years later, H. J. Coolidge (1933) made a detailed anatomical inspection of a pregnant female specimen collected in Lukolela. In his bulky monograph published in the *American Journal of Physical Anthropology*, Coolidge asserted that this second type of chimpanzee merited separate species status for the following reasons: the anatomical measurements obtained for this second type, the pygmy chimpanzee, were smaller than the smallest value for any other subspecies of chimpanzee; the pygmy chimpanzee displayed a total of 19 infantile features; its behavior clearly differed from that of the common chimpanzee; and its intellect possibly surpassed that of the common chimpanzee.

Coolidge also stated that it was possible that *Pan paniscus* is a true paedomorphic species, in which the adult form retains juvenile characters, and is closer to the common ancestor of chimpanzee and man than is any other existing chimpanzee species or subspecies. According to the "biogenetic law" that ontogeny recapitulates phylogeny, the lower the degree of specialization (or the more primitive or juvenile the form) and the more highly specialized the mature (or adult) form that is achieved by a lineage, the closer the less specialized form is thought to be to the ancestral condition in that lineage. At least one fact supporting Coolidge's hypothesis is evident even to non-scientists. That is, the skull of a pygmy chimpanzee is the exact likeness of the skull of *Australopithecus*, the oldest known ancestor of man (Fig. 5).

For a long time after R. A. Dart (1925) discovered the skull of the "Taung Child," which he called *Australopithecus*, the scientific community did not recognize this ape-man as an early member of the human line. Indeed, the fact that several special features of the pygmy chimpanzee were very similar to those of *Australopithecus* provided one reason for questioning the identity of *Australopithecus* as a hominid. That these similarities provided Dart's opponents with a powerful weapon was unfortunate, and, ironically, one of the antagonists was none other than Ernst Schwarz, who had described the type specimen of *Pan paniscus*.

R. Broom (1925) became one of Dart's supporters, but even after the discovery of a fossil of an adult *Australopithecus* at Sterkfontein, Schwarz (1936) voiced opposition. He said (p. 969), "The small size, the dome-shaped forehead, the shortness of the face which is responsible for the crowding together of the teeth, are the same in both [*Australopithecus* and *Pan satyrus paniscus*]."

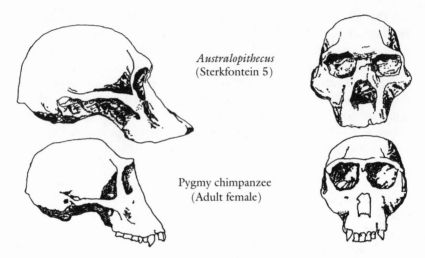

Australopithecus
(Sterkfontein 5)

Pygmy chimpanzee
(Adult female)

Figure 5. External features of the skulls of an *Australopithecus* and a pygmy chimpanzee (Redrawn from Zihlman et al., 1978; by permission of Plenum Press)

Broom (1936) asserted that the absence of a space between the lateral incisors of the upper jaw and the canines was the single most important feature proving the affinity of *Australopithecus* and man. Schwarz (1936), however, countered that while spaces between these teeth exist in the gorilla and common chimpanzee, they may be insignificant in the pygmy chimpanzee. Finally, Schwarz (1936: 969) declared, "The fact that Dr. Broom's specimen does not represent an ancestral form of the hominoid line does not detract from the extreme value of the discovery. It is to be hoped that he will be able to continue his researches, and to elucidate the history of anthropoids in Africa, an undertaking quite as important and interesting as that of the ancestors of man himself." In this way, all of the humanlike special features of the pygmy chimpanzee ended up being used as evidence of the "ape-ness" of *Australopithecus* instead of being regarded as evidence for the affinity of the pygmy chimpanzee and man. In those days, this was, perhaps, unavoidable.

In 1933, Coolidge advocated full species status for the pygmy chimpanzee. In an essay on chimpanzee classification, in the following year, however, Schwarz opposed that viewpoint and reaffirmed limiting the pygmy chimpanzee to the subspecies level. In 1954, E. Tratz and H. Heck put forth a new view. They raised the status of the pygmy chimpanzee to a genus separate from the common chimpanzee and advocated calling it

Bonobo paniscus. Insufficient new facts were raised, however, to be convincing. Thus, these are the three different historical views about the classificatory position of the pygmy chimpanzee: a subspecies, a species, or a separate genus. Today, the overwhelming majority of researchers use the species name *Pan paniscus* for the pygmy chimpanzee, but this seems merely a "custom," and its proper classificatory position remains unresolved.

In the 1960's, comparative research on the pygmy chimpanzee and the common chimpanzee advanced rapidly. That research was not limited to morphology and anatomy, but extended to physiology, biochemistry, and genetics. The work of, among others, Adrienne L. Zihlman of the University of California and her colleagues had a great impact on the study of pygmy chimpanzees in the latter half of the 1970's. In 1978, Zihlman et al. studied the comparative morphology of the pygmy chimpanzee and *Australopithecus*. They found that the two species are very similar in body build, and that the differences relate to the mode of locomotion (*Australopithecus* is a biped; the pygmy chimpanzee is a quadruped). According to the molecular clock estimates, the divergence of the chimpanzee lineage and the human lineage was 4 to 5 mya, and Zihlman et al. proposed that the pygmy chimpanzee be a model for the common ancestor of man and apes.

The immediate response to Zihlman et al.'s theory was silence. In academia, a reaction is not instantaneous or reflexive, as in boxing. It takes time to become armed with the necessary evidence to attack another's theory or to defend one's own. Not until 1981 did representatives from paleontology, paleoanthropology, physical anthropology, primatology, and every possible related field attempt, in unison, to discredit it. B. M. Latimer et al. (1981) said there was absolutely no foundation for asserting that the pygmy chimpanzee, among the extant apes, is the most similar to *Australopithecus afarensis*. Steven C. Johnson (1981) stated that the small size of the pygmy chimpanzee was no more than an island effect caused by isolation, just as forest elephants and antelope are smaller in size than the savanna forms. Others argued that the pygmy chimpanzee was not primitive, but instead had become specialized. Like bees protecting a hive, these researchers, one by one, directed stings at the research results and methods of Zihlman et al., and criticized their comparative research as completely meaningless.

A conclusion, however, has not yet been reached in the dispute over the genealogical status of the pygmy chimpanzee. Like a swinging pendulum that has not yet settled, the phylogenetic position of *Pan paniscus*

A young male pygmy chimpanzee (*Pan paniscus*). His hair is black. He has relatively small ears, and long cheek hairs jut out on both sides of his face. Pygmy chimpanzees are more agile in the trees than common chimpanzees.

oscillates between, at one extreme, Zihlman et al.'s theory and the opposite viewpoint. We need time and a better understanding of the pygmy chimpanzee to determine its place in nature.

Special Features of Pygmy Chimpanzees

Here I shall present a rough sketch of the special features of the pygmy chimpanzee.

External appearance. In Table 2, I compare the external appearance of both species of chimpanzee. Six of the nine distinctive features high-

Kajabara, a common chimpanzee (*Pan troglodytes schwein-furthii*) in Tanzania, East Africa. Facial hair is short, and the ears are prominent. The body build is more robust. (Junichiro Itani)

light the unique juvenile characteristics of the pygmy chimpanzee: the body hair is predominantly black (3); the ears are relatively small (4); the head does not go bald (5); a tail tuft persists into adulthood (6); inter-digital webbing occurs between the second and third toes (7); and the forehead is round and high and the supraorbital torus is small (9). The color of the face and palms of the hands of the common chimpanzee is light until juvenile age and may be an "infantile" feature.

Skull. Douglas L. Cramer (1977) compared the skulls of pygmy and common chimpanzees (Fig. 6), and some of the results are summarized

]27[

Table 2. Comparison of the external appearance of the pygmy chimpanzee and the common chimpanzee

Pygmy chimpanzees	Common chimpanzees
1. The skin on the face and extremities stays dark for life, but around the eyes and lips it is a pale flesh color.	From infancy to the juvenile stage, the skin is a pale flesh color; in adulthood, it often appears freckled or mottled.
2. Long cheek hair protrudes on both sides.	No cheek hair resembling that found in pygmy chimpanzees is present.
3. The body hair is black, though in exceptional cases, it may be brown.	The body hair is black, turning light brown or gray in adulthood.
4. The ears are relatively small and are positioned close to the head.	The ears are relatively large and jut out from the sides of the head.
5. The hair on the head is dense; adults do not become bald with age.	Some adults become bald on the forehead.
6. There is a spot at the base of the spinal column where there is no hair. Below that point, there is a short, white tail tuft that persists even into adulthood and is never conspicuous.	The tail tuft, very conspicuous in infancy, disappears as the individual matures.
7. About half of all individuals display webbed skin between the second and third toes of the feet.	Few individuals have interdigital membranes (eight cases out of a sample of 109, according to A. H. Schultz, 1956).
8. The body is relatively small and of slender build. The limbs are relatively long. The forelimbs and hindlimbs are of almost equal length, and when a pygmy chimpanzee stands quadrupedally, the body is horizontal.	The body is relatively large and of stout build. The forelimbs are longer than the hindlimbs, and when a common chimpanzee stands quadrupedally, the body is inclined, higher at the front.
9. The forehead is relatively high and round; the brow ridge is slight.	The forehead is relatively low; starting at the brow ridge, the growth of the facial bones is marked.

in Table 3. In the main dimensions, the pygmy chimpanzee is smaller than the common chimpanzee. Values only slightly less than those of the common chimpanzee are found in vault length, vault height, postorbital breadth, and cranial capacity. A large difference between the two exists, however, in the mandibular length and maximum palatal length. This means that the pygmy chimpanzee's cranium is relatively globular and rounded; in the common chimpanzee, facial prognathism provides a striking difference. This difference in the degree of facial prognathism may also be regarded as showing the juvenilization of the pygmy chimpanzee.

Cramer (1977) found neither sagittal nor nuchal crests in his pygmy chimpanzee specimens, but observed sagittal crests in 4% and nuchal crests in a higher percentage of common chimpanzee specimens. Accord-

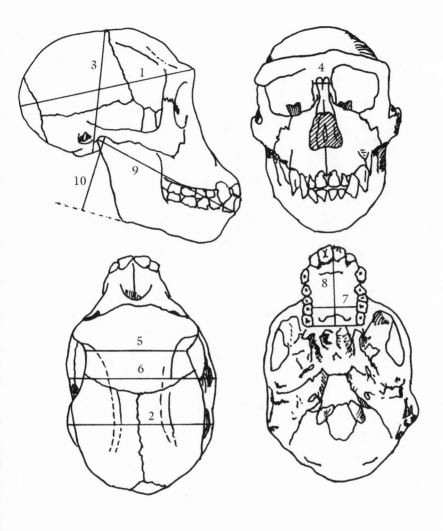

1 Total vault length	6 Bizygomatic breadth
2 Total skull breadth	7 Palatal breadth
3 Vault height	8 Maximum palatal length
4 Interorbital breadth	9 Mandibular length
5 Postorbital breadth	10 Mandibular height

Figure 6. Comparison of craniofacial features of pygmy chimpanzees and common chimpanzees (Cramer, 1977; by permission of S. Karger AG, Basel)

Table 3. Comparison of the craniofacial measurements
of the pygmy chimpanzee (PC) and the common chimpanzee (CC)
(Cramer, 1977; by permission of S. Karger AG, Basel)

Measurement	PC	PC/CC	CC
Vault length	122.8 mm	0.91	135.3 mm
Vault breadth	105.8 mm	0.87	121.5 mm
Vault height	85.4 mm	0.96	89.3 mm
Interorbital breadth	14.5 mm	0.73	20.0 mm
Postorbital breadth	65.6 mm	0.94	69.7 mm
Bizygomatic breadth	110.9 mm	0.89	124.9 mm
Palatal breadth	51.5 mm	0.86	59.7 mm
Palatal length	58.1 mm	0.76	76.3 mm
Mandibular breadth	93.4 mm	0.89	105.2 mm
Mandibular length	102.4 mm	0.79	130.0 mm
Mandibular height	55.4 mm	0.83	66.9 mm
Cranial capacity	350 cm^3	0.90	389 cm^3

Table 4. Comparison of the anatomy (in mm) and body weight (in kg)
of the pygmy chimpanzee (PC) and the common chimpanzee (CC)
(Zihlman and Cramer, 1978, by permission of S. Karger AG, Basel;
Horn, 1979; Rahm, 1967; Wrangham, personal communication)

	PC (both sexes)	PC/CC	CC (female)
BODY			
clavicle length	105	0.86	122
scapula length	138	0.99	140
scapula breadth	72	0.95	76
innominate length	253	0.98	259
iliac breadth	97	0.84	116
sacrum breadth	63	0.93	67.5
FORELIMBS			
humerus length	285	0.997	286
radius length	262	0.992	264
ulna length	274	0.986	278
HINDLIMBS			
femur length	293	1.043	281
tibia length	242	1.039	233
fibula length	218	1.000	218

NOTE: Pygmy chimpanzees: wild and captive, average 35.5 kg (25–48 kg, 13 wild and five captive, Zihlman and Cramer, 1978); captive (Frankfurt Zoo), 42–46 kg (>20 years old, male), 35 kg (<20 years old, female) (Horn, 1979)
Common chimpanzees: wild, average 42 kg (25–50 kg, Rahm, 1967); wild, Gombe Park, average 35 kg (39.5 kg, male; 30 kg, female; Wrangham, personal communication)

ing to W. G. Kinzey (1984), because the dentition of the pygmy chimpanzee and the common chimpanzee shares several advanced and primitive traits, we cannot determine which is closest to humans or which is most primitive.

Body weight and special features of the postcranium. The average body weight of a pygmy chimpanzee is 84.5% that of a common chimpanzee, although overlap is seen. Zoo-reared pygmy chimpanzees can become misleadingly larger in body weight than wild common chimpanzees, supporting the argument that the pygmy chimpanzee was never a pygmy (dwarf) type.

The postcranium (the whole skeleton excluding the skull) of the pygmy chimpanzee is, in general, smaller than that of the common chimpanzee (Table 4). The hindlimb bones are an exception, however, with the pygmy chimpanzee's being longer. Bones in the forelimbs of the pygmy chimpanzee are shorter, but the difference is insignificant. Thus, the length of the four limbs in the pygmy chimpanzee is relatively long compared with the common chimpanzee. The hindlimbs are especially long,

The ratio of forelimb to hindlimb in pygmy chimpanzees is close to 1 : 1.

] 31 [

and the ratio of forelimb to hindlimb is close to 1 : 1. It is possible that the ratio of the length of the limbs relative to the trunk, which is a function of size, is a primitive characteristic retained from proto-apes (Shea, 1984).

Sexual variation/dimorphism. The pygmy chimpanzee is the least sexually dimorphic primate. Cramer (1977) compared the measurements of 20 cranial variables of male and female pygmy chimpanzees, and found little difference: the female's mean value is 98.6% of the male's. Moreover, in four cranial variables, females show an even higher value than males. In the common chimpanzee, the mean value of females is 97% that of males, and in only one variable does the female's value exceed that of males. According to D. C. Johanson (1974), the teeth are also similar between the sexes, and differences occur only in the canines.

Body build. A. L. Zihlman (1984) compared the body build of apes, *Australopithecus*, and modern humans. The relative weight of the upper limbs in both species of chimpanzee is 15.8% of total body weight, which is greater than in *Australopithecus* (12%). The relative weight of the hindlimbs of the pygmy chimpanzee is 24.2%, however, which is larger than

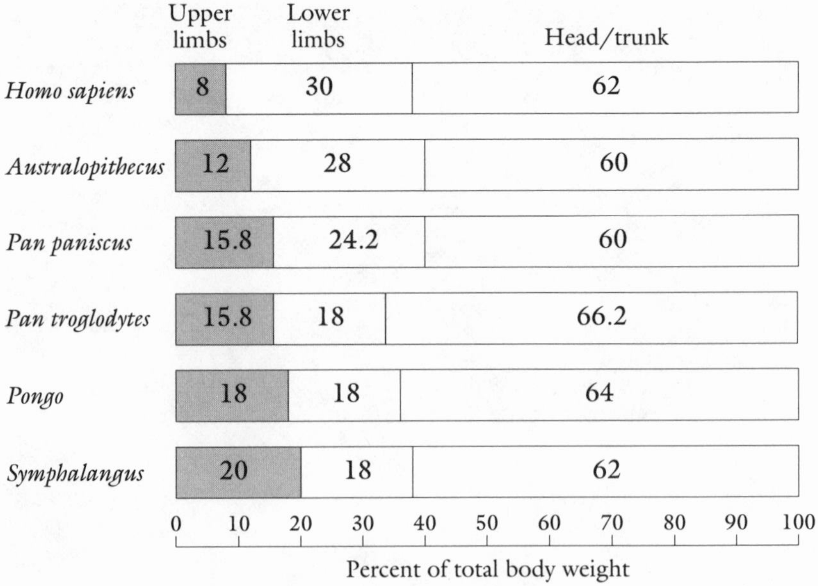

	Upper limbs	Lower limbs	Head/trunk
Homo sapiens	8	30	62
Australopithecus	12	28	60
Pan paniscus	15.8	24.2	60
Pan troglodytes	15.8	18	66.2
Pongo	18	18	64
Symphalangus	20	18	62

0 10 20 30 40 50 60 70 80 90 100

Percent of total body weight

Figure 7. Distribution of body weight to body segments in five female hominoids. The body weight distribution for *Australopithecus* was estimated (Redrawn from Zihlman, 1984; by permission of Plenum Press)

Table 5. Comparison of blood groups of the pygmy chimpanzee
and the common chimpanzee (after Socha, 1984)

Pygmy chimpanzees	Common chimpanzees
A-B-O Blood Group System	
A type and O type (A antigen is indistinguishable from human A_1 type)	A and O (A_1 and A_2 subtypes of the A antigen; both differ from the human counterparts)
M-N Blood Group System	
Only M (M antigen differs from both the human and the common chimpanzee forms)	M_1 type and MN type
V-A-B-D Blood Group System	
Only one v.D type has been found, and it was seen in all research subjects (14 individuals); v.D type occurs at a frequency of only 1.4%	Extremely variable; 16 types reported
Rh Blood Group System	
In general, the two species resemble each other, but the reaction of Rh_0 antigen is lower than that of common chimpanzees	Resembles that of pygmy chimpanzees, but the reaction of the Rh_0 antigen is stronger
R-C-E-F Blood Group System	
All individuals examined (14) were $R_{ab}CE$ type	Twenty-two blood-group types have been defined; of these, 20 were actually observed. Five of the blood-group types that were irregular were rare, one of them $R_{ab}CE$. Of 570 subjects examined, there was not one common chimpanzee of $R_{ab}CE$ type

that of the common chimpanzee (18%) and approaches that of *Australopithecus* (28%) (Fig. 7). If an ape resembling the pygmy chimpanzee were the ancestor of *Australopithecus*, the morphological-anatomical gap between a quadrupedal ape and the ancestor of man would disappear, according to Zihlman.

Blood types. Human blood types were discovered in 1901, and research on the antigens of red blood cells of apes began after 1960. The latest research on the blood groups of the pygmy chimpanzee is from W. W. Socha (1984), whose findings are outlined in Table 5. According to this research, the pygmy chimpanzee differs from the common chimpanzee in five blood-type systems. According to Socha, a difference of this magnitude indicates different genera, not just species. By contrast, the blood types within the three subspecies of the common chimpanzee display only very small differences.

Time of divergence. According to the molecular clock, the chimpanzee, gorilla, and human diverged almost simultaneously about 5 mya. Thus, the chimpanzee genus *Pan* has a history of five million years. According to the analysis of mitochondrial DNA, which makes substitutions ten times faster than nuclear DNA, the genus *Pan* has accumulated about a 6% change during this time. Because the common and the pygmy chimpanzee share 4% of this 6%, and accumulated the remaining 2% after they diverged (Sarich, 1984), the divergence of the pygmy chimpanzee from the common chimpanzee may have occurred about 1.5 mya. Using the same method, many think that the divergence time of the three subspecies of common chimpanzee is probably 1 mya.

The pygmy chimpanzee differs from the common chimpanzee in so many features of morphology and physiology that I think they should be considered distinct species. Later, I will report large differences also in their ecology, society, and behavior. Nevertheless, some primatologists, such as Arthur D. Horn (1979), contend that the pygmy chimpanzee is no more than a subspecies of the common chimpanzee, *Pan troglodytes*. If this contention is accepted, the existing three subspecies of the common chimpanzee may have to be abolished, because the differences between the pygmy and the common chimpanzee exceed those among the three races of the common chimpanzee. Also, some primatologists believe that the pygmy chimpanzee compared to the common chimpanzee is paedomorphic, or neotenous (retains juvenile characteristics after reaching sexual maturity). As a general tendency, this is probably correct. However, the question of whether or not the pygmy chimpanzee is closer to the common ancestor of the chimpanzee lineage is a matter of deciding how to evaluate neoteny.

A Brief History of Field Research

As mentioned previously, field research on apes is intended to help us explain evolution by exploring the origins of human behavior and society. Here I want to review briefly the history of chimpanzee field research in Africa.

Foremost among laboratory researchers on apes was Robert M.

Yerkes, who also was the first to sense the need to conduct behavioral research on apes in a natural state. In 1920, he sent two young researchers, H. W. Nissen and H. C. Bingham, to West Africa to investigate the chimpanzee and the gorilla, respectively; he also sent C. R. Carpenter to Asia to carry out investigations of gibbons. These three men made many important observations in natural settings, but then political conditions changed and field research on apes entered a long blank period.

About 1960, ape research in Africa recommenced, almost simultaneously begun by Japan, the USA, Britain, and Holland. From Japan, under the leadership of Kinji Imanishi, Masao Kawai and Hiroki Mizuhara went to the border area of Rwanda and Zaire and made short-term studies of gorillas. At the same time, the American George B. Schaller was to do a long-term study of gorillas in the same region. As a result, the Japanese team shifted to observing chimpanzees in Tanzania under the leadership of Junichiro Itani. After setting up Kabogo, Kasakati, and Kasoge bases, Itani's group, gradually and after much difficulty, succeeded in provisioning at Kasoge Base in 1966. Toshisada Nishida who is the founder of the Kasoge Base, along with many young researchers, has obtained almost all the research findings about wild common chimpanzees by Japanese at this research site.

The Dutchman Adriaan Kortlandt (1962) was the first person to succeed in closely investigating wild chimpanzees in the Beni Forest on the eastern border of Zaire. He also surveyed a wide area throughout Africa and collected a great deal of valuable, detailed information about the distribution of chimpanzees. According to Kortlandt and his successors, the site of Guinea's Bossou region, which he established and maintained, is even now one of the influential research sites for observing the common chimpanzee. Yukimaru Sugiyama from Japan also conducts research at Bossou.

The English woman Jane Goodall was the first to succeed in provisioning and habituating wild chimpanzees. When Nishida began his investigations at Kasoge, Goodall was already in Gombe National Park (today, a game reserve). Here she made many valuable discoveries that astonished the world. For example, she observed chimpanzees termite-fishing and tool-making, meat-eating, and food-sharing. In 1968, Goodall published a monograph about the common chimpanzees of Gombe Park that has been an important source of information for chimpanzee researchers and has retained its brilliance to this day. Since the latter half of the 1960's, many young American researchers have gone to Gombe Park, which has become a British-American site of cooperative research.

About the same time that Goodall was at Gombe Park, a British couple, Vernon and Frances Reynolds, undertook research on common chimpanzees in the Budongo Forest of Uganda. The chimpanzees of this forest were found not only in very high density but they were partially habituated, so that even without provisioning, the Reynoldses could observe them in the wild state. The Budongo Forest was a very important site, where, later, Yukimaru Sugiyama and Akira Suzuki also conducted research. After Idi Amin became the dictator of Uganda, however, investigations became difficult and were suspended.

In addition to those already mentioned, there are long-term field research sites for common chimpanzees in Rio Muni (Jones and Sabater-Pi, 1971) and in Senegal's Mount Assirik (McGrew, Baldwin, and Tutin, 1981). If sites providing short-term investigations are included, the number is very large, and the number of researchers probably exceeds 80.

Field research on the pygmy chimpanzee got a late start, in the early 1970's, ten years after work had begun on the common chimpanzee. In the 1960's, there was no possibility of pursuing investigations because of the long, continuous political agitation in the Congo. At that time we did not completely understand the conditions of the pygmy chimpanzee in the Zaire basin, and even feared that they might be extinct. Toshisada Nishida was a pioneer in field research on the pygmy chimpanzee. Nishida (1972) briefly surveyed the region of the west bank of Lake Tumba, and reported that pygmy chimpanzees were not yet extinct there. They were, however, cornered by critical conditions caused by human pressures.

In 1973, based on Nishida's advice, I undertook a wide survey in Zaire. Although this survey proposed to investigate the distribution of the pygmy chimpanzee (results presented in the next chapter), the primary goal was to discover a suitable site for continuous long-term research. There were three criteria for the selection of my research site. First, chimpanzees must be abundant. Second, human pressure must be small; specifically, villagers did not kill or eat chimpanzees. Third, transportation must be convenient and the procurement of supplies easy. On the basis of nearly five months of surveying, I chose two sites: Wamba, at the southern end of the zone of Djolu, and Yalosidi, at the southern end of the zone of Ikela.

By chance, also in 1973, the Irishman Noel Badrian and his wife began a wide survey with the same goal in mind. After driving all the way from the Commonwealth of South Africa to the Zaire Basin, they chose Lomako, in the Befale region, as a suitable site.

Also at the same time, two parties from the USA went to Zaire to investigate pygmy chimpanzees. One party, from the Yerkes Primate Research Center, had as its priority the organized capture of pygmy chimpanzees. Their goal was to gather specimens quickly, not to observe, quietly and patiently, the animals in the field. Ignoring the protests of many primate researchers who wanted to protect the pygmy chimpanzee, the Yerkes group executed their plan, captured a number of live pygmy chimpanzees, and bred them for research in the laboratory.

Another party from the USA was led by Arthur D. Horn of Yale University (currently at the University of Colorado). Using Nishida's then current report, Horn attempted to investigate pygmy chimpanzees on the west bank of Lake Tumba for nearly two years. However, because the density of pygmy chimpanzees was very low, he had to switch to observing monkeys during the course of his study. During his two-year investigation, direct observations of pygmy chimpanzees were recorded only a few times.

Since 1974, I have been able to direct a long-term research project on pygmy chimpanzees, funded by Monbusho, or the Ministry of Education, Science, and Culture, Japan, which has acknowledged the scientific value of pygmy chimpanzees. For the first three years, we decided to pursue parallel research projects divided between two sites, Wamba and Yalosidi. Members of our team took turns: at the Wamba site, Suehisa Kuroda, Kano, then Kano again, and at Yalosidi, Kano, Shigeo Uehara, followed by Kohji Kitamura. At both sites, we did our best to habituate the chimpanzees through artificial feeding. However, the chimpanzees at Wamba were the first to be successfully provisioned, and after 1977 we discontinued research at Yalosidi, pouring all our energies into Wamba.

Thereafter, excluding a gap of one year (1985), research at Wamba has been carried out every year. Alone I visited Wamba four times, in 1977, 1978, 1981, and 1984, and during this time span, provisioning and habituation advanced on a large scale. Now, all the individuals in one out of five groups of pygmy chimpanzees inhabiting the forests of Wamba have been identified, and many individuals of three of the remaining four groups have been identified.

I personally have accumulated about 1,800 hours of direct observations, but the total observation time is considerably more, owing to additional observations made by Suehisa Kuroda, Kohji Kitamura, Takeshi Furuichi, Akio Mori, Genichi Itani, and others. In this way, Wamba has developed favorably as a pygmy chimpanzee field site, but not without problems. Among them is a relationship between the scientific grant period for research funding from Monbusho and observations being limited to a period from September to March. Because most of our observations cover only the seasons between September and March, an important problem for the future is how to fill the blank period from April to August.

At present, Lomako is the only pygmy chimpanzee research site other than Wamba. Noel Badrian has a position at the State University of New York, and American researchers are centered at that university, continuing research under the supervision of Randall L. Susman. Because these researchers avoided provisioning, they were apparently unable to observe their subjects easily. Recently, however, the pygmy chimpanzees at Lomako have become gradually habituated; the research is finally advancing, and the researchers have begun to recognize individuals.

CHAPTER TWO

Distribution

The Historical Record

From the discovery of the pygmy chimpanzee in 1928 until about 1933, there was a succession of locality reports, but afterward not much was made of these findings. There is, however, unpublished material from the Belgian colonial period at the Central African Scientific Research Institute (IRSAC) located at Lwiro on the west bank of Lake Kivu, which was the center of research in the natural sciences in Zaire, Rwanda, and Burundi. IRSAC possesses carefully collected reports about the locality. These locality records are pointed out on Map 2, and when I set out to survey in 1973, they constituted all the known reports of pygmy chimpanzees. Generally, we understand their distribution to be in the area between the Zaire River in the north and the Kasai-Sankuru River in the south.

There are, however, a few records of pygmy chimpanzees from the north bank of the Zaire River. The records of IRSAC imply that pygmy chimpanzees exist at Makanza and Lisala, towns on the north bank of the middle Zaire River, and on the north bank of Lake Kivu. According to A. Urbain and P. Rode (1940), there are reports of pygmy chimpanzees living in the basin of the upper Sangha River, a tributary of the north bank of the Zaire River (see Map 2, only the lower portion is drawn).

On the basis of my survey, it is doubtful that these exceptional reports are reliable. Although records of the past provided primary leads at the start of investigating the distribution, usually when I actually made an on-site visit, suitable conditions for chimpanzees did not exist. In general, if I asked people who lived at the sites (mostly central towns), many answered that they had not seen chimpanzees in the forests since they were children.

Many animal specimens collected in the past probably were bought

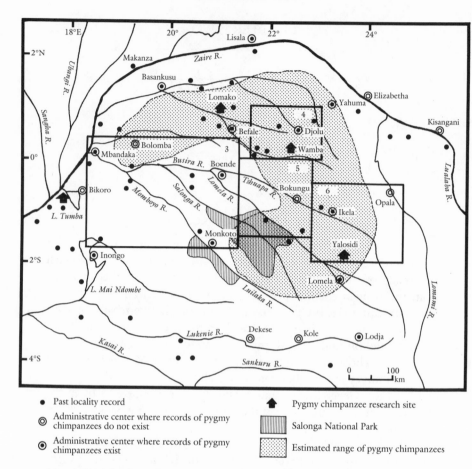

Map 2. Distribution of pygmy chimpanzees (adapted from Kano, 1984a; by permission of S. Karger AG, Basel) This map is based upon information gathered by the author in 1973. The present distribution of pygmy chimpanzees is probably smaller. The large rectangles indicate the boundaries of Maps 3–6.

from local inhabitants, and such purchases identified a locality as a regional collection and distribution area. It must have been at these local, central towns where collectors in the past could easily find their specimens. Antelopes, monkeys, birds, and various other animals were brought to local trading centers from distant villages, where they were sold in the market, not only as pets but as food. Chimpanzees also were eaten in many regions. Therefore, it is no wonder that many of the locality records came from such central towns as Mbandaka, Boende, and Monkoto. If the specimens were purchased in the market, most of these locality reports are unreliable.

For these reasons, we should probably remove Lisala and Makanza from the producing area. The information from Lake Kivu is perhaps also mistaken. Quite strangely, a report from the neighboring Republic of the Congo (former French Congo) on the upper Sangha River says that a French company officer captured an adolescent pygmy chimpanzee in 1939 (Urbain and Rode, 1940). It was undoubtedly a pygmy form, but I find it hard to believe that the upper Sangha forest was its natal place because the river was a barrier. The report from the Sangha River must be reconfirmed.

I am convinced that on the north side of the Zaire River, there are no pygmy chimpanzees. At least, they do not have a natural habitat there. On the other hand, a southern limit exists, although some people believe that the pygmy chimpanzee once extended as far south as Angola (Reynolds, 1967), on the basis of Tulp's description in 1641. Tulp's description, however, is not necessarily evidence that Angola was formerly home to the pygmy chimpanzee for the previously mentioned reason—the specimens could have been imported. The chimpanzees brought to Europe from Angola after Tulp's specimen were unquestionably a "common" form. This shows that the Portuguese colony Angola was at that time one of the large commercial centers on the African west coast.

In general, it is probably reasonable to assume that the geographic range of the pygmy chimpanzee is located within the boundary of the southern bank of the Zaire River, east to west for 900 km, and fills the oblong area, from north to south, for 600 km.

Research in Recent Decades

In recent years, only the Badrians and I have investigated pygmy chimpanzees over a fairly wide area of their range. None of us covered the whole territory because our main objective was to look for a candidate site for intensive studies. The Badrians traveled by Land Rover along the west coast of Lake Tanganyika from Zambia and headed north, entering Kisangani from the state of Kivu. Following a tip, they entered Befale, but the details of that excursion are not published.

My trip was not as long as the Badrians', but since I walked, the survey in the region between the Zaire and Luilaka Rivers was a little more careful. I was unable, however, to get to the central and eastern regions of the Ikela zone, to the eastern regions, or to the zones of Yafuma and Basankusu. Having relied on past records and having ended up on a trip that was completely in vain (in which I had plunged headlong from Monkoto to Salonga National Park) I learned from experience the value of gathering extensive information beforehand and relying on tips from inquiries.

Plantations exist here and there in the Zaire Forest. (Where the cultivation of commercial crops is large-scale, the size of the smallest one is about 1–2 km². In the Zaire Basin, palm oil, coffee, and rubber are the principal crops.) The head offices and branch offices of these plantations are located at district towns, such as Mbandaka and Boende. I first questioned Belgians, who at that time were still living there in large numbers. The information that I obtained from them, being mostly secondhand, was not always accurate. However, because many of them hold more than one plantation job and travel around, I used their information to get a rough idea of a wide territory. In a moderate-size town, I was always able to obtain information of similar quality from Catholic priests and school teachers.

Next, I actually went to the sites, met the villagers, and began to inquire whether or not chimpanzees were in the neighboring forest. Here the villagers, except for a few, did not understand the French words for pygmy chimpanzee, *chimpanzé nain*. The Lingala word *mokomboso* also was strange-sounding to them.

In historical records, several local names for pygmy chimpanzee have been reported. In addition to the earlier mentioned *bonobo*, there is *edja* or *edja mbanda* (Kundu tribe dialect), *elya mana* (a dialect of a riverine tract from Lomela to Tshuapa Rivers) (Callewaert, 1930), *elya* (Lake Tumba coast), and others.

The people who inhabit the Zaire Basin zone are a large tribe called the Mongo, which is divided into more than 40 branch tribes who speak their own unique dialects. For example, the inhabitants of the Wamba region are a branch of Mongo called the Gandu tribe. The dialects resemble each other closely, and people from different tribes can understand each other, even though they cannot speak each other's dialects. There are many words in common, such as animal and plant names. This is true for "pygmy chimpanzee," although only one word, *elya* (pl. *bilya*), was used throughout the entire region that I traversed. In the Ikela region, there are people who pronounce it with a little accent, *eza* or *edja*. From Djolu, in the Bongandanga region, *engombe* (pl. *bingombe*) is also used. Never, however, did I hear "bonobo," which is being used in

A young hunter south of the Zaire River, with bow and
arrows and a recently killed redtail monkey.

some academic journals. "Bonobo" is a term introduced by E. Tratz and
M. Heck (1954) as a native name for the pygmy chimpanzee, but it does
not seem to have general usage in the principal domain of the pygmy
chimpanzee.

What was fortunate for me was that the people on the south bank of
the Zaire River knew the forest well. In contrast, according to Colin M.
Turnbull (1968), the Bantu farmers of the Ituri on the north bank of the
Zaire River hate and fear the forest and seldom enter it. Whereas they are
said to know very little about the forest, the Mongo tribe of people on
the south bank, also Bantu farmers, have a profound knowledge of the
animals and plants of the forest. They cultivate cassava (*Manihot utilis-*

Ngandu women carrying baskets loaded with leaves and firewood.

sima) as their staple food along with some vegetables, but they obtain a considerable portion of supplementary food from the forest. Especially, almost all their animal protein sources are from wild animals—fish, mammals, birds, amphibians, reptiles, and insects. The number of animal species collected may exceed 100, even excluding fish. Products from the forest are not just used as foods; hundreds of kinds of plants are used as medicines and as materials for houses and tools. The men are not only capable hunters; they are also excellent natural historians. Because each individual is self-sufficient and there is no social stratification or division of labor, almost all men in their prime have an excellent knowledge of nature.

My first question was whether or not chimpanzees were in the nearby forests. I expected an answer that was 100% correct, but since the villagers' daily sphere of activity centers around their home, with a radius generally of 5 to 6 km, information was reliable only within that range. One of the discoverers of the common chimpanzee in Tanzania, C. H. B. Grant (1946: 110) wrote, "Native information is, as I know too well, most unreliable and they will tell a white man any tale they think will please him." Among our Mongo people, however, none fabricated stories about the presence or absence of chimpanzees. Information was obtained from 103 villages that *bilya* were present. Out of these, I tried to verify the reports from 31 villages by entering the forest myself. If I saw some trace or sign of chimpanzees, as was true for 30 villages, I tended to believe the accuracy of their reports. I also went to the one remaining village, but it rained heavily and, disappointed, I had to turn back after only a 20-minute survey. I felt certain that some trace of chimpanzees would have been found if there had been more time.

The results of my inquiries are shown in Maps 3–6. In the literature, the eastern boundary of pygmy chimpanzee distribution is tentatively described as the Lomami River, but I was unable to verify it. Father François of the Yaloya Catholic Church in the southern part of the Ikela zone, who is an adventurer and fan of hunting, has twice gone upstream on the

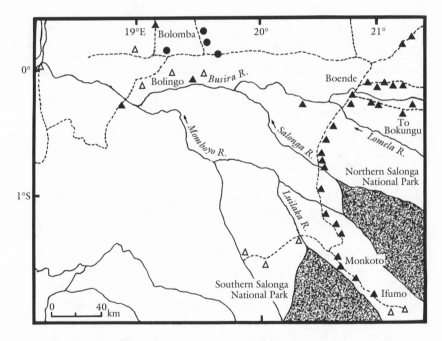

● Presence of pygmy chimpanzees confirmed by on-the-spot survey by author

△ Presence of pygmy chimpanzees confirmed by miscellaneous sources
(e.g., missionaries, teachers, plantation directors)

▲ Absence of pygmy chimpanzees confirmed by inquiry and survey

⌁ Road

Map 3. Western distribution of pygmy chimpanzees, based on Map 2 (adapted from Kano, 1984a; by permission of S. Karger AG, Basel)

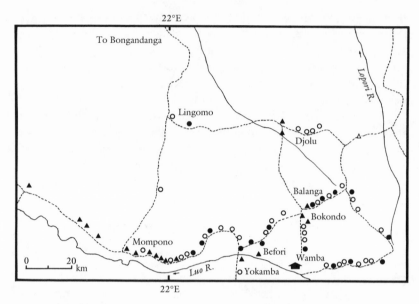

22°E

To Bongandanga

Lingomo

Djolu

Lopori R.

Balanga

Bokondo

Mompono

0   20 km

Befori

Wamba

Luo R.

Yokamba

22°E

● Presence of pygmy chimpanzees confirmed
by on-the-spot survey by author

○ Presence of pygmy chimpanzees confirmed
by local inhabitants

△ Presence of pygmy chimpanzees confirmed
by miscellaneous sources (e.g., missionaries,
teachers, plantation directors)

▲ Absence of pygmy chimpanzees confirmed
by inquiry and survey

⬗ Pygmy chimpanzee research site

⌒ Road

Map 4. Northern distribution of pygmy chimpanzees, based on Map 2 (adapted from Kano, 1984a; by permission of S. Karger AG, Basel)

Tunbenga River in eastern Ikela, but he said he did not see traces of chimpanzees. Likewise, Hiroaki Sato, who twice conducted anthropological research at Moma on the eastern border of Ikela, has not seen chimpanzees. He says that if they do exist there, they are probably extremely scarce. Perhaps the pygmy chimpanzee did not reach the Lomami River. The Badrians (pers. comm.) are of the same opinion about this. Although I am uncertain about the eastern boundary, if I were to draw a distribution line fairly boldly, the size would come to about 135,000 km². This encompasses only the northern territory between the Zaire and Luilaka rivers.

I was not able to investigate the southern region from Kasai to the basin of the Sankuru River, but this may be separated, in the main, from the northern part of pygmy chimpanzee distribution. Once I took a trip from Lomela to Loja, but the people along the road said there were

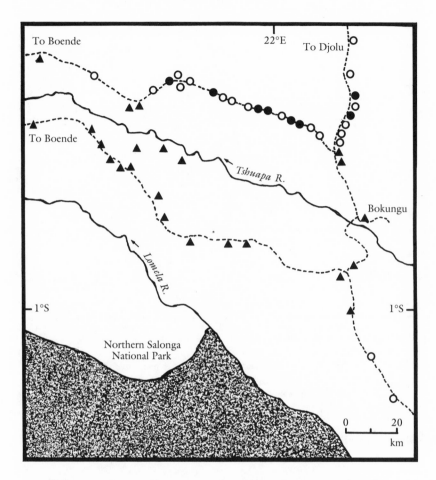

To Boende

22°E

To Djolu

To Boende

Tshuapa R.

Bokungu

Lomela R.

1°S

1°S

Northern Salonga
National Park

0   20
km

● Presence of pygmy chimpanzees confirmed by on-the-spot survey by author

○ Presence of pygmy chimpanzees confirmed by local inhabitants

▲ Absence of pygmy chimpanzees confirmed by inquiry and survey

⌇ Road

Map 5. Central distribution of pygmy chimpanzees, based on Map 2 (adapted
from Kano, 1984a; by permission of S. Karger AG, Basel)

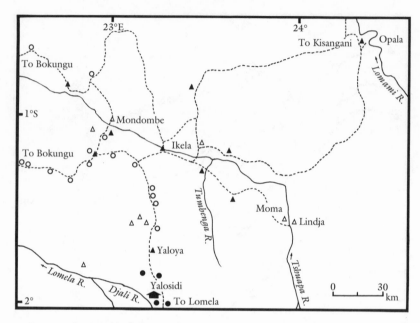

● Presence of pygmy chimpanzees confirmed
  by on-the-spot survey by author

○ Presence of pygmy chimpanzees confirmed
  by local inhabitants

△ Presence of pygmy chimpanzees confirmed
  by miscellaneous sources (e.g., missionaries,
  teachers, plantation directors)

▲ Absence of pygmy chimpanzees confirmed
  by inquiry and survey

◂ Pygmy chimpanzee research site

⌇ Road

Map 6. Eastern distribution of pygmy chimpanzees, based on Map 2 (adapted from Kano, 1984a; by permission of S. Karger AG, Basel)

no chimpanzees. The northern area of distribution may connect with the southern one somewhere close to the Zaire River between the Busira River and Lake Tumba. The reports of A. D. Horn (1980) and Nishida (1972), however, show that the territory where Lake Tumba levee touches that part is narrow and the density of pygmy chimpanzees is extremely low.

In addition to the northern area of distribution, which is covered with the vast Zairean forest, rain forests exist in fragments within the grassy plains as far south as the region of Kasai (Map 7). Recently, however, there have been no reports of pygmy chimpanzees from the southern part of their distribution, and we fear that they may be extinct there. If they do exist, they are probably barely clinging to life in these isolated

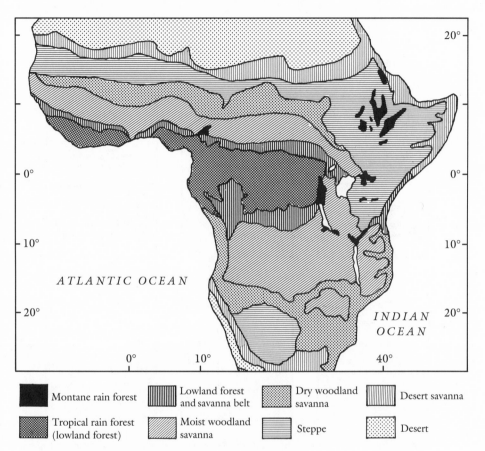

Montane rain forest

Tropical rain forest (lowland forest)

Lowland forest and savanna belt

Moist woodland savanna

Dry woodland savanna

Steppe

Desert savanna

Desert

Map 7. Vegetation of Africa.

forests, and they probably survive only in small numbers. It is safe to say that the majority of wild pygmy chimpanzees inhabit the northern area. The size of the whole range of distribution of the pygmy chimpanzee probably does not exceed 200,000 km².

Ecological Conditions

In the pygmy chimpanzee's area of distribution, climate and topography are invariable. Throughout the whole territory, the average monthly air temperature from the highest to the lowest is 30° C and 20° C, respectively; therefore, hardly any monthly variation is noticeable (Fig. 8).

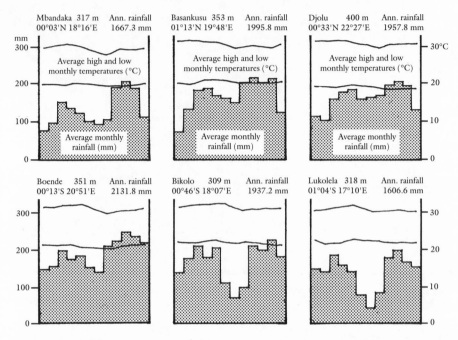

Figure 8. Rainfall, temperature, and altitude above sea level within the pygmy chimpanzees' region of distribution (Vuanza and Grabbe, 1975).

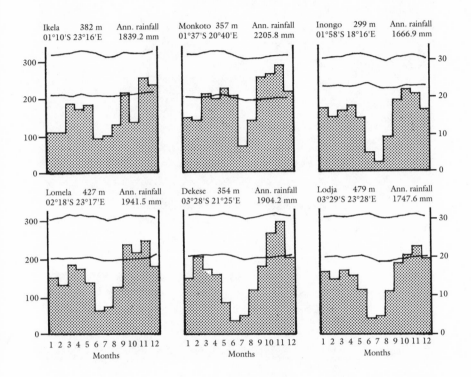

Ikela   382 m   Ann. rainfall
01°10'S 23°16'E   1839.2 mm

Monkoto 357 m   Ann. rainfall
01°37'S 20°40'E   2205.8 mm

Inongo   299 m   Ann. rainfall
01°58'S 18°16'E   1666.9 mm

Lomela   427 m   Ann. rainfall
02°18'S 23°17'E   1941.5 mm

Dekese   354 m   Ann. rainfall
03°28'S 21°25'E   1904.2 mm

Lodja   479 m   Ann. rainfall
03°29'S 23°28'E   1747.6 mm

Months

Dry primary forest.

Swamp forest.

In the northern part of their distribution, the annual rainfall is between 1,600 and 2,000 mm and, in general, is evenly distributed among months. From the equator southward, rainfall clearly follows the pattern of the southern hemisphere, with the minimum amount of rain falling between the months of June and July. There are more than six months when the amount of rainfall per month exceeds 100 mm and only two to three months of less than 25 mm of rainfall, satisfying the conditions for the formation of a tropical rain forest.

The altitude is 479 m at Lodja, the highest point, and 299 m at the lowest point, Inongo. The difference in altitude is just 180 m in the southern part of their distribution, and in the northern part, the difference is even smaller. There the whole range in altitude is from 350 to 400 m above sea level. Observations in the field attest to the monotonous character of geographic features. In general, even river cuts in the flat areas are shallow, seldom forming deep valleys or gullies. There is some relief to the land, but very few high places from which you can overlook the forest. Next I shall describe the chimpanzee's most important floristic environment.

In the Zaire forest, what could be called primary forest is dry forest

Forest that has been cleared for cultivation.

Table 6. Common plant species in the forest (listed in order of abundance)

Vernacular name	Scientific name	Family name
Abundant tree species in dry primary forest:		
bofili	*Scorodophloeus zenkeri*	Caesalpiniaceae
bolingo	*Anonidium mannii*	Annonaceae
bosefe	*Garcinia epunctata*	Guttiferae
langa	*Brachystegia laurentii*	Caesalpiniaceae
bokumbo	*Leonardoxa romii*	Caesalpiniaceae
bosole	*Angylocalyx pynaertii*	Papilionoideae
bondenge	*Duvigneaudia inopinata*	Euphorbiaceae
bolinda	*Polyalthia suaveolens*	Annonaceae
bottende	*Pancovia laurentii*	Sapindaceae
bombongo	*Gilbertiodendron dewevrei*	Caesalpiniaceae
elimilimi	*Dialium pachyphyllum*	Caesalpiniaceae
bokolombe	*Staudtia stipitata*	Myristicaceae
lambo	*Cleistanthus mildbraedii*	Euphorbiaceae
beko	*Celtis mildbraedii*	Ulmaceae
bimba	*Paramacrolobium coeruleum*	Caesalpiniaceae
loyombiya	*Tessmannia africana*	Caesalpiniaceae
esousou	*Santiria trimera*	Burseraceae
botuna	*Cynometra hankei*	Caesalpiniaceae
Abundant trees in the swamp forest:		
bosenge emoke	*Uapaca heudelotii*	Euphorbiaceae
waka mobali	*Baikiaea insignis*	Caesalpiniaceae
waka mwasi	*Julbernardia seretii*	Caesalpiniaceae
botongo	*Xylopia phloiodora*	Annonaceae
lambo	*Cleistanthus mildbraedii*	Euphorbiaceae
bosole	*Drypetes gossweileri*	Euphorbiaceae
ilo ya mai	*Diospyros alboflavescens*	Ebenaceae
beembe	*Eriocoelum petiolare*	Sapindaceae
bolilo	*Raphia sese*	Palmae
embanje	*Heisteria parvifolia*	Olacaceae
lileke	*Pandanus* sp.	Pandanaceae
bokola	*Coelocaryon preussii*	Myristicaceae
Abundant tree species in old secondary forest:		
bondenge	*Duvigneaudia inopinata*	Euphorbiaceae
bopola	*Alchornea floribunda*	Euphorbiaceae
lomuma	*Anthonotha macrophylla*	Caesalpiniaceae
bottende	*Pancovia laurentii*	Sapindaceae
bofili	*Scorodophloeus zenkeri*	Caesalpiniaceae
bokana	*Panda oleosa*	Pandaceae
bolimo	*Ochthocosmus africanus*	Ixonanthaceae
bofumbo	*Grewia pinnatifida*	Tiliaceae
bosenge	*Uapaca guineensis*	Euphorbiaceae
bokongo	*Antrocaryon micraster*	Anacardiaceae
bosulu	*Pterocarpus soyauxii*	Papilionoideae
bokungu	*Piptadeniastrum africanum*	Mimosaceae

Table 6. Continued

Vernacular name	Scientific name	Family name
Abundant trees in young secondary forest:		
bonyanga	*Croton haumanianus*	Euphorbiaceae
bombambo	*Musanga smithii*	Moraceae
liyamba	*Albizzia gummifera*	Mimosoideae
bila	*Elaeis guineensis*	Palmae
bosakesake	*Caloncoba welwitschii*	Flacourtiaceae
bokomokomo	*Barteria nigritana*	Flacourtiaceae
yenge bosala	*Macaranga saccifera*	Euphorbiaceae
yenge yokuluha	*Macaranga monandra*	Euphorbiaceae
yenge bochacha	*Macaranga spinosa*	Euphorbiaceae
bolungu sele	*Fagara lemairei*	Rutaceae
nyangosake	*Lindackeria dentata*	Flacourtiaceae
bokungu	*Piptadeniastrum africanum*	Mimosaceae

and swamp forest, but in addition to this, there is also secondary vegetation. Secondary vegetation emerges where there were cultivated fields and dwellings. Soon after a field is abandoned, it is replaced by secondary bush, which does not contain any medium to tall trees. The secondary bush then gradually develops into secondary forest.

Swamp forest, dry (primary) forest, and secondary forest all have unique tree types, and several representative species from each are placed together in Table 6. For example, in the primary forest, there are many bean trees of the suborder Caesalpiniaceae, whereas in the secondary forest, members of the Euphorbiaceae, Mimosaceae, Flacourtiaceae, and others are present. For convenience, I divided forests into the first stratum (tall tree layer, $\geq$25 m), second stratum (medium tree layer, 15−25 m), third stratum (low tree layer, 5−15 m), and fourth stratum (shrubs and herbaceous layer, 0−5 m). Each stratum, however, is usually continuous with its neighbors, and there are several different ways to divide the forests into strata.

Almost all rivers and streams are fringed with swamp forests of various sizes. Alongside large rivers, large-scale swamp forests may reach several kilometers in width. In flood times, they become forests floating in water, and the people navigate their boats within the groves and angle for fish. The tree known to villagers as *bosenge emoke* (scientific name: *Uapaca hendelotii*) and many other kinds of trees in the swamp forest have prop roots like the arms of an octopus. There are also trees that have aerial roots. Because the soil is loose, some swamp trees do not stand straight,

Bosenge tree (*Uapaca hendelotii*), showing prop roots.

and there are many large, inclined trees. Many trees are crowded together, with intertwined roots; often the trees appear to support each other. If you walk on the floor of the swamp forest when it is dry, you will feel the matlike surface roll under you. Because the height of the trees is relatively low, seldom exceeding 30 m, the swamp forest is usually bright; however, at some locations, there are also totally dark groves with a closed canopy.

The type of forest that I tentatively call "dry" in this book is also very humid. It differs from the swamp forest in having solid soil instead of a muddy floor with scattered water pools. Because the tree crown overlaps at many levels, the forest floor is generally dark, and the undergrowth is relatively sparse. In the highest tree layer, gigantic trees (emergents) reaching as much as 40–50 m are also seen. The dry forest has the most complicated floristic composition, and pure stands seldom occur. The dry forest is included in many forest types that differ from each other in tree density and composition, but this will not be touched on here.

In the land of the Mongo, fields are made by cutting a clearing in the dry forest. Therefore, all secondary forests derive ultimately from the dry forest, but because the primary forest trees are generally much larger and harder to cut down, villagers have a tendency to avoid clearing new sections of primary forest. Instead, in most villages at present, slash-and-burn horticulture has a closed rotation only in secondary vegetation. Secondary vegetation goes from fields (2–3 years) → secondary bush (3, 4–5, 6 years) → young secondary forest (6, 7–9 years, 10 years) → fields.

In secondary bush, which appears first in abandoned fields, African ginger and herbs of the family Marantaceae (3–5 m high) and shrubs at the 5-m level cover the area. A few trees that remained uncut at the time a field was opened may be scattered among those thick herbs and shrubs. Trees grow fast in secondary bush, which is replaced by young secondary forest where the trees reach 15–20 m. The canopy is usually open in 6–7 years.

What is called old secondary forest is secondary forest that has been abandoned for a long period, perhaps for special reasons like emigration. The tall tree stratum becomes dense, but the middle and low tree strata are sparse. The herbs (Marantaceae) grow thick on the forest floor or twine around the trunks of tall trees. In old secondary forest, tree species characteristic of young secondary forest and dry primary forest mix. Two or three tree species are most abundantly seen in old secondary forest (Table 6). Under certain unknown conditions, one type of dry primary forest is similar to old secondary forest, with dense herbs (Marantaceae) and sparse tall trees. It differs from normal dry primary forest in physiognomy, so for convenience, I classify it as old secondary forest in this book.

Each of the above-mentioned forest types has its respective unique composition and appearance. Moreover, within a particular type, regional differences are exceedingly few. The trees often seen in primary forest are seen in this kind of forest no matter where you go. The same is true for the other forest types. From one region to another, however, the proportion of forest types differs. For example, there is a great tendency for the proportion of swamp forest to decrease toward the east and on upstream to the upper tributaries of the Zaire River. Conversely, the proportion of dry primary forest increases from west to east, and secondary forests broaden near large towns and plantations.

Population Size and Density

Judging from my experiences in both Tanzania and Zaire, the chance of directly encountering chimpanzees in a broad regional survey is extremely rare. In my five-month survey in 1973, I made a long journey in search of chimpanzees in the Zaire basin, but I observed them only 14 times. Therefore, in estimating population density, I had to rely on the information of local residents or other indirect evidence.

To estimate the density of pygmy chimpanzees, I first divided the evidence gathered in the forest into four categories: (1) direct observations (also including vocalizations), (2) nests, (3) remains of food, and (4) other indirect evidence.

(1) Evidence from direct observations is free from error, but the small amount that was gathered is a problem. (2) Nests provide good evidence. With practice, one can quickly recognize chimpanzee nests. Even after they have become fairly old, it is impossible to mistake them for, for example, bird nests. Because they are made at least once per day and last an average of about four months, chimpanzee nests are easy to count. (3) Food remains also can be gathered in abundance, but have two defects as evidence. It is difficult in some cases, first, to distinguish pygmy chimpanzee food remains from those of other monkeys and, second, to record the amount objectively. (4) In the class of other indirect evidence, I include footprints, feces, snapped-off dragged branches, and so on. I did not collect many kinds of exact data in this class; it is also difficult to record the amounts accurately. Data in categories (1) and (2) constitute the main evidence, whereas those in (3) and (4) are secondary.

On the basis of the 33 places I investigated by entering the forest, I tried to estimate the relative density. I concluded from those findings that Wamba followed by Yalosidi are the regions where the density of pygmy chimpanzees is highest. In later investigations (Kano and Mulavwa, 1984), the density of chimpanzees at Wamba was estimated at 1.7 individuals/km². On the basis of this estimate, I concluded from looking at the relative quantity of evidence at the 33 locations including Wamba that the average value of the density of pygmy chimpanzees was 0.7 individuals/km².

The calculations have some problems, however. Within the range I drew as their northern area of distribution, 99 localities and 40 nonlocalities were intermingled. That is, there were lots of holes in this region. When you take this into consideration, the whole regional distribution comes to an average of 0.4 individuals/km². Consequently, using the above-mentioned method I estimated 54,000 pygmy chimpanzees

(135,000 km² × 0.4 individuals/km²) in the northern territory. This is a completely arbitrary method of calculation, but I have not been able to find another more appropriate method of deriving definite values. Moreover, this figure is likely to be an overestimation. I went to spot-inspect the forest only where a high density of chimpanzees was expected on the basis of the reports of villagers and the appearance of the forest.

Because almost all pygmy chimpanzees inhabit the northern territory, the total number of pygmy chimpanzees in the world probably does not exceed 100,000 individuals. My opinion is that the total is at most about 50,000 individuals. Moreover, breeding opportunities appear to be suppressed because of the extremely low density throughout most of the region. There is the strong possibility that they may continue to survive in extremely difficult conditions.

Factors Determining Distribution and Density

The Zaire and Kasai rivers, respectively, are established as the northern and southern limits of distribution of the pygmy chimpanzee. These enormous rivers, which span at their widest points 2 to 3 km, have existed for a long time and probably continue to stand as constant geographic barriers to pygmy chimpanzees. Distributional limits are set not only by these major rivers, however. Along the several tributaries of the southern banks of the Zaire River, rivers having a width of more than 50 m even at their upper parts are quite common. These probably also obstruct the interchange of members of individual groups.

The Mongo people say that pygmy chimpanzees can swim. Certainly, pygmy chimpanzees are not as extremely fearful of water as common chimpanzees, and they enter the swamp forest and feed in streams 20–30 cm deep. I once saw footprints of a pygmy chimpanzee who moments before had forded a river 60 cm deep and 3 m wide, probably crossing it in a bipedal walk. However, I find the assertion that chimpanzees swim to be far from believable.

The pygmy chimpanzees of Wamba could have been completely separated from those of Ilongo for decades or even centuries by the Luo River, even though they could undoubtedly hear and, on occasion, see each other. Although the pygmy chimpanzees of the Wamba region seem to be continuously distributed, the whole population is actually cut up into several isolated populations. This phenomenon, seen in my map of their range, will offer interesting materials for future study of variation in the ecology, society, behavior, and morphology of the pygmy chimpanzee.

Although the northern and southern limits of distribution are set by rivers, the eastern boundary is different. If the Badrians' and my guess is correct, the distribution there is not obstructed by rivers. Pygmy chimpanzees disappear somewhere in the forests of the Ikela zone between the Tshuapa River and the Lomami River. If so, "vegetation" this time becomes the primary factor limiting distribution.

According to C. Evrard (1968), a professor of Botany at Kinshasa University, the Zaire Basin can be divided into three vegetation zones: the Western swamp forest zone, the Eastern dry forest zone, and, in between, the mixed forest zone. According to these divisions, the greater part of pygmy chimpanzee distribution in the northern region occurs in what Evrard calls the swamp forest zone, and the eastern part of the Ikela can barely be placed in the western part of Evrard's mixed forest zone. I must point out that my vegetation classification differs from Evrard's. What Evrard classified as swamp forest, I divide again into dry forest and swamp forest. That is, my dry forest is not identical to Evrard's dry forest.

Even though there are no clear geographic barriers, the density of pygmy chimpanzees gradually decreases within Evrard's mixed forest zone as pygmy chimpanzees approach the eastern dry forest zone. The pygmy chimpanzee is perhaps an ape that has adapted to the most humid part of the tropical rain forest. On the other hand, if you consider my vegetation zone divisions only (dry forest, swamp forest, and secondary forest), the pygmy chimpanzee seems to rely most strongly on dry forest. Within the northern part of their distribution, proceeding from east to west (that is, as the proportion of swamp forest gets larger), the density of pygmy chimpanzees seems to decrease. As I will discuss later, the major portion of the pygmy chimpanzee's staple food is taken from the dry forest.

People rival rivers and vegetation as one of the main factors influencing the distribution and density of pygmy chimpanzees. In many regions of the Mongo tribe, pygmy chimpanzees are hunted for food. Where human hunting pressures are high, as along the west shore of Lake Tumba, it is possible that chimpanzees have decreased in density. For that region, I tried to compare the amount of evidence of pygmy chimpanzees recorded per unit of time in hunting areas and non-hunting areas. Contrary to expectations, I did not observe any difference.

Of interest, however, is that people who live in areas where chimpanzees are used for food always take notice of them. They remember well where something was eaten or where they saw footprints and nests; when asked for guidance in the forest, they would go directly to the

An orphaned infant whose mother had been killed and eaten.

place. On the other hand, people who do not eat chimpanzees usually are completely disinterested in them; they must start from scratch to search for them when they enter the forest. This difference can bias data gathered during short-term research.

Next, I considered whether the villagers' information about the "density" of pygmy chimpanzees was useful or not. The majority of reports are from villages where people responded with the greatest exaggeration, saying they are "innumerable" or "our forest is just teeming with them." Therefore, reports that say "there are many" are unreliable. Among these, however, there were also reports that "I think there are some in spots scattered around" or "I think you will see them but it takes one week." I think we can rely on the "low density reports," recognized as those that say these things, and if we compare the circumstances to the exaggerated reports, it is clear that many low density reports were from villages where hunting pressures are high.

Even if humans do not kill chimpanzees directly, their effect through the indirect destruction of vegetation is great. Although the small secondary forests formed by villages cause changes in vegetation, conditions may still be acceptable for chimpanzees. But the building of large-scale plantations or the formation of regional towns may completely strip the forest that is their source of life.

Human pressure influences not only the density of pygmy chimpanzees but also their distribution. Within the boundaries of their distribution, there are countless pockets where chimpanzees do not exist, and these are clearly the result of direct and indirect influences of man. Also present are vast empty zones in the middle and lower basin of the Tshuapa River and southward. These may be related to strong "human pressure" in concert with a low original density of chimpanzees there. In fact, among the people of this region, many assert that "if there are bilya (pygmy chimpanzees), we will eat them." The ancestors of these people may have hunted chimpanzees.

The distribution of the pygmy chimpanzee may have been long influenced by strong hunting pressures beginning as much as 700 years ago, ever since Bantu farming tribes migrated into the Congo forest zone. Because pygmy chimpanzees are large-bodied and have the habit of making a lot of noise when they arrive at feeding sites and before going to sleep, they can be located easily, relative to other animals, and are an easy hunting target.

Comparison with Common Chimpanzees

The common chimpanzee has a long, extensive range from Senegal on the West African coast, across the northern part of the equatorial rainforest zone of the Zaire River, to both sides of Lake Tanganyika in East Africa, and as far south as 8° S (Fig. 5). The length of this belt is about 6,000 km, and the average width is 500 km. The range is an estimated 10 to 15 times that of the pygmy chimpanzee, probably providing the greatest difference in distribution patterns between the two species. On the basis of its range of distribution, the common chimpanzee is the most successful species of great ape (see Map 1).

The three subspecies of chimpanzee are arranged along this long strip of distribution in order, starting from the west: West African chimpanzee (*Pan troglodytes verus*), Tschego (*P. t. troglodytes*), and East African chimpanzee (*P. t. schweinfurthii*). The line dividing the West African and Tschego chimpanzees is the Niger River, whereas the Ubangi River separates the Tschego and East African chimpanzees. In this way, the separation of chimpanzees, whether as subspecies or species, seems to be largely caused by rivers. Among them, the Zaire River, which separates the pygmy chimpanzee from the common chimpanzee, forms certainly the oldest geographical barrier.

Within the large distribution zone, the habitat diversity of the common chimpanzees is impressive. First, they have an altitude range from 0 m to 3,000 m above sea level (Kortlandt, 1967), which is very close to that of humans. Second, climatic conditions are also diverse. The West African chimpanzee exists in humid areas, for example, at Conakry (Guinea), where the average annual rainfall reaches 4,350 mm (range, 3,211–5,470 mm) (de Bournonville, 1967). The same West African chimpanzee also survives in the Mount Assirik district of Senegal where average annual rainfall is just 954 mm (range, 824–1,224 mm) (McGrew, Baldwin, and Tutin, 1981). Moreover, the slopes on the shore of Mount Cameroon, which is in the distribution zone of the Tschego, is said to have an annual rainfall of 12 m. At the other extreme, in the dry Ugalla hill zone that forms the eastern boundary line for the East African chimpanzee, the average is presumed to be 900 mm or less.

Reflecting these diverse topographies and climatic conditions, the floristic environment of the common chimpanzee is complex. They have successfully advanced into many areas, from the tropical lowland forests extending from West Africa through Central Africa, to the mountain forests that have developed in Cameroon and East Africa's Western Great Rift zone, to the deciduous woodland-savanna belt in the boundary zone of their distribution. Among savanna-type vegetations, woodland is the most moist vegetation. Trees and shrubs 10–15 m high are found in considerable density, but the canopy does not grow thick enough to overlap. Because the woodland floor is covered with grasses of the family Gramineae, a glance indicates that it is distinct from a tropical rain forest, which has no grasses but rather herbs and shrubs. Also called savanna forest, woodland is the general term.

The common chimpanzee probably depends primarily on the rain forest because the bulk of its range occurs there. In East Africa, however, it is found 450 km south of the eastern boundary of the equatorial rain-forest belt. In those regions, for the most part, the floristic composition is a mosaic of savanna vegetation that comes from woodland, savanna, and grassland. The tropical rain-forest vegetation exists in nothing more than a form called "riverine forest," which occurs in a thin line along the river. The broken hills of Ugalla, which form the eastern limits of the distribution of the East African common chimpanzee, are the driest among the suitable habitats. There, even the riverine forest is scattered and discontinuous, occupying no more than a scant 0.7% (15 km²) of the total area of Ugalla (2,700 km²) (Kano, 1972). Adriaan Kortlandt believes that in the old days when human pressure was not great, the common chimpanzee may have lived farther north than it does now. They probably returned to the tropical rain-forest belt in response to hunting pressure and destruction of vegetation.

Pygmy chimpanzees on the forest floor.

Pygmy chimpanzees in the trees in dry primary forest. The primary habitat of pygmy chimpanzees is tropical rain forest.

Ndilo, a young female common chimpanzee, on the ground, in Tanzania. (Junichiro Itani)

Common chimpanzees of Mahale, Tanzania, in the trees. The habitat of common chimpanzees is drier than that of pygmy chimpanzees. (Michael Huffman)

In this way, the common chimpanzee is a "euryoecious species" that occupies a wide niche. By contrast, the pygmy chimpanzee is limited to a habitat in the wet part of the tropical lowland rain forest and may be a relatively "stenoecious species." This difference may be a reflection of environmental differences existing during the past 1.5 million years since the two species diverged from their common ancestor.

Recent paleo-geologic evidence about African Pleistocene environments suggests that, in the driest periods during the Pleistocene, most of the equatorial rain forest was replaced by savanna vegetation. In such periods, tropical rain forests remained in isolated patches in the high mountains such as Ruwenzori and Cameroon. The most recent such event occurred close to 19,000 years ago. At that time, the Kalahari Desert extended to the north and jutted out nearly to the Zaire estuary (Hamilton, 1976).

At the other extreme, during the maximum pluvials of the Pleistocene, a huge lake was formed in the Zaire Basin. This lake is called "Bushira," after one of the largest tributaries that gathers the waters of the Tshuapa, Lomela, Salonga, and other rivers in the central Zaire basin.

The region on the north side of the Zaire River extending east and west was probably more strongly affected by the latest violent climatic fluctuation than by the complex topography. Formation of the great rift valleys and orogenic activity, which had been advancing in an east-west direction, were additional influences. The ancestors of the common chimpanzee, playing it both ways, must have survived alternately in the advancing savanna and in the forest.

By contrast, I wonder if the ancestors of the pygmy chimpanzee were left in a relatively stable forest environment. To the north, east, and west, pygmy chimpanzees were obstructed by the Zaire and Lualaba rivers. To broaden their distribution, they could advance only to the south. The vegetation in the southern boundary region of their present distribution is a mosaic of tropical rain forest and grassland, and the intermediate woodland and savanna are not seen there. In the past, if intermediate vegetation (transition zone) had difficulty growing in that region, and if the ancestors of pygmy chimpanzees tried to escape from the forest, they should have immediately entered grasslands.

Going from forest directly to grassland is a great fall from what ought to be called a "vegetation precipice." In addition, a large geographic barrier called the Kasai River blocks the way farther to the south. Under these floristic conditions, the pygmy chimpanzee was unable to acquire a eurytopic nature like that of the common chimpanzee and had

] 65 [

to accept its stenoecious nature, which was locked into the tropical lowland forest.

We do not understand very well the conditions in the Zaire Basin
during the driest periods of the Pleistocene, but geographers (Kadomura, 1980) think that possibly almost all the forests ended up being replaced by savanna vegetation there. I disagree with this conclusion, however, because the relatively stenoecious pygmy chimpanzees could not
have weathered the dry period and continued to survive in that region.
Even in the recent driest times, after the water of Lake Bushira dried up,
a refuge forest of sufficient size must have remained to rear enough of
their ancestors.

CHAPTER THREE

Social Groups and Social Patterns

Four Patterns

Nonhuman primate social units can be roughly divided into four types. The first is a society in which each individual lives separately. Males and females meet and have sexual relationships only in the mating season. Junichiro Itani (1972) calls such a society, made up of only these kinds of independent or solitary individuals, an "elementary society." Although common in prosimians, it applies only to orangutans among the higher primates.

In general, social units other than elementary societies take a form called a "social group." These are divided into (1) monogamous, (2) uni-male, and (3) multi-male groups.

The monogamous group, also called a pair-type society, consists of one male and one female who have a sexual relationship and their off-spring. As each offspring approaches sexual maturity, it is repulsed by the parent of the same sex and leaves the group. Young males and females who separate from their group encounter each other by chance and form new pairs. The pair-type society occurs in prosimians and New World monkeys, but is rare among Old World monkeys. In apes, only the gibbon has this form of society.

In a monogamous group, a pair is formed anew each generation, and we cannot say if it is patrilineal or matrilineal. The uni-male and multi-male groups, however, are divided into two types: a matrilineal one, in which the group is maintained by males leaving the natal group, and a non-matrilineal one. Uni-male groups are often seen in the super-family Cercopithecoidea, and the hanuman langur may be considered a representative of that type.

The average hanuman langur group is made up of one male, five fe-males, and eight young individuals (Sugiyama, 1965). The male offspring leave the group before reaching sexual maturity and enter all-male groups that are nomadic and independent of the mixed groups. A male in a mixed group is often challenged by males from within these all-male groups. If the male in the mixed group wins, he remains in that group, but if he is defeated, his position is sought after, and he may have to join an all-male group. Because the average duration of a male's reign is 4½ to 5 years, he successfully avoids entering into sexual relations with his own daughter.

The hamadryas baboon and the gelada baboon have a group structure, maintained by male transfer, that is a transition form between uni-male and multi-male group structure. Among the hamadryas baboons, several harem groups range and feed separately throughout the day, but at night, they gather at one resting place and form very large sleeping congregations. The harem groups, however, do not intermix spatially. Gelada baboons, on the other hand, do not break up into smaller units, even during the daytime; they always appear to act as a large multi-male group. Careful observation, however, reveals several harem groups gathered together, and they, like those of the hamadryas baboon, do not mix spatially.

The multi-male type of society, represented by the savanna baboons and macaques (Japanese monkeys, rhesus monkeys, etc.), may have expanded. A typical group is organized with the highest-ranking males, females, and their offspring occupying the central part of the group; lower-ranking males crowd around the periphery. In Japanese monkeys, the majority of males without the distinction of high rank give up membership in the natal group after reaching sexual maturity, usually within five years (Koyama, 1977). The males that remain in the group rarely copulate with mothers or sisters. In other words, the Cercopithecoidea have a matrilineal social group in which males leave the group.

The gorilla and chimpanzee take a uni-male and multi-male form, respectively, and they differ in their maintenance of social structure. The gorilla has a social unit comprised of a patriarchal male and several females with their offspring. As the male infants grow, it is not uncommon for them temporarily to take the form of a multi-male group. Sooner or later, however, the male offspring leave the natal group, and the group returns to a uni-male form. In gorilla society, both males and females give up membership in their natal group. The females emigrate to another group, but males do not. Males do approach other groups, however, and lure away (kidnap) young females that become future consorts. This contact serves as the germ of new unit groups. The society of gibbons is similar to that of gorillas in that both males and females give up membership in the group.

The orangutan lives a solitary existence. A single male or female ranges alone, and the only time a male and a female get together is when they happen to encounter each other and mate. A female who has given birth lives with her infant until the infant reaches 4–5 years of age. Then it leaves its mother and enters an independent life of its own.

Chimpanzee society also has unique features. Among these is that some individuals come and go, freely forming and leaving groups. Although hamadryas baboons have what is called a fixed fission-fusion unit, consisting of a male and several females and offspring, chimpanzee fission-fusion units are not so clearly fixed. This social flexibility is the most distinctive feature of chimpanzee society.

Chimpanzee researchers discovered this fission-fusion phenomenon in the early years and called this temporary gathering a "temporary association" (Goodall, 1965) or a "band" (Reynolds and Reynolds, 1965). For some time they had thought that the social unit that in other primates is like a "troop" did not exist in common chimpanzee society; that associations between individuals who temporarily gather was not limited even by geographical barriers; and, that spatially, they spread out everywhere. Goodall's early paper asserts that in chimpanzees, there are no permanent ties beyond that of mother and child. V. Reynolds (1966) proposed further that human society originated in an open group like that of the chimpanzee.

T. Nishida (1968) discovered that common chimpanzee society is made up of groups that contain numerous individuals but are limited in size and that members from different groups do not gather, even temporarily. He named these distinct groups, "unit groups." He found at Kasoge, his study site in western Tanzania, that when chimpanzees from the south (M-group) arrived, the provisioned chimpanzees (K-group) avoided them, moved away from the feeding place, and finally headed north without ever joining the new arrivals. The fission and fusion of chimpanzees occurs only within the framework of the "unit group."

Although at first Western primatologists did not readily accept the existence of unit groups, no one now doubts that the unit group is the social unit of the common chimpanzee. Nevertheless, Western scientists use a different term, "community," instead of "unit group." Further research has shown that females, especially young females, transfer from unit group to unit group, groups that are closed to males but open to females. Thus, males spend a lifetime in the group into which they were born, revealing that common chimpanzee society has a patrilocal form, a rare phenomenon in Old World primates. These two characteristics—a fission-fusion nature and a patrilocal residence form—within the framework of the unit group are the unique social features of the common chimpanzee.

Soon after starting to investigate the pygmy chimpanzee, we confirmed the existence of structures corresponding to unit groups in its society, as well as the occurrence of fission and fusion within them. As with the common chimpanzee, and accompanying progress in habituation, we began to observe the custom of young nulliparous females emigrating between unit groups. This is a basic social characteristic that the pygmy chimpanzee shares with the common chimpanzee.

When we compare social structures in more detail, however, we discern various differences between the two species. Below, I summarize superficially what we know about the differences in the social characteristics of the pygmy chimpanzee and the common chimpanzee. In the following chapters, I shall consider these in greater depth. It is also important to realize that we cannot appraise social differences if we omit considering the environment, the animals' methods of exploiting it, and their sexual and social behaviors, because social differences are strongly related to differences in means of survival and reproduction.

The Unit Group and Fission and Fusion

Here, I shall clarify the fission-fusion nature of the chimpanzee unit group. Unit groups within the genus *Pan* are basically social units consisting of males, females, and immature individuals (juveniles, infants). These constituent members do not move in unison, as do the savanna baboon and macaques. Usually, they are divided into several smaller groups. Goodall, the discoverer of the chimpanzee's fission-fusion nature, called the above-mentioned small groups "temporary associations," but recently, the term "party" suggested by Y. Sugiyama (1969) is being applied. From now on, most textbooks will probably use the term "party."

The party does not have fixed constituent members. For example, if individuals A-B form one party, when that pair joins another party and becomes A-B-C-D-E, the individuals have the flexibility, depending on the opportunities, to divide into two other partylike forms, such as B-D and A-C-E. The constituent members of a unit group do not form parties only in this way. Many become temporarily solitary, but the unit group serves to bring all of them together. Because females, especially young nulliparous ones, come and go between groups, the unit group is difficult to describe. Rarely do we observe all the constituent members of a unit group assembled into one party.

The form of fission and fusion in unit groups is important to decide. For example, how many chimpanzees are required to form a party (size)? For how long does membership remain stable (continuity)? Who associates with whom frequently (fission-fusion unit)? Below, I shall explore these questions with respect to the pygmy chimpanzee.

According to my reports of the provisioned E-group at Wamba up to 1979, the size of the parties encountered was from 1 to 40 individuals within the five unit groups; most parties were from 6 to 35 individuals (Map 8, Table 7). The party membership of E-group changed slowly. Membership in a party commonly is stable for two to three weeks, and sometimes even more than a month. Because of this slow rate of change, if the first researchers had been able to undertake only short-term research, they may have concluded that pygmy chimpanzee groups are like troops of Japanese monkeys and baboons.

A complete pygmy chimpanzee party is comprised of adult males, adult females, and offspring (adolescents, juveniles, and infants; see Table 8). According to my observations, up to 96% of the groups were this kind of mixed party. The ratio of males to females deviates greatly from 1:1 in a few parties. The ratio of adults and adolescents to juveniles and infants is about 3:1. Males are sometimes solitary and, occasionally, several will form an all-male party. These last, at most, only one to two days. No matter how they change in size, pygmy chimpanzee parties usually include a balance of males, females, and offspring. Uni-sexual parties and mother-offspring parties are rare.

In pygmy chimpanzee parties, which are not always mixed parties, the proportion of estrous females can reach an extremely high rate, as much as 98%. Therefore, I thought that the flexibility of pygmy chimpanzee society was reflected in its ability to provide both the functions of reproduction and child-rearing in every phase of fission and fusion. Certainly, if the group does not lose the ability to breed when it divides into parties, it will probably increase its chances of survival even in a severe environment.

The smallest mixed party consists of one male, one female, and their offspring, which is, in form, identical with the human family. I once thought that the prototype of the human family was the pygmy chimpanzee form of society. I also thought, when this kind of group was put in a harsh environment (such as savanna), it split into male-female pairs

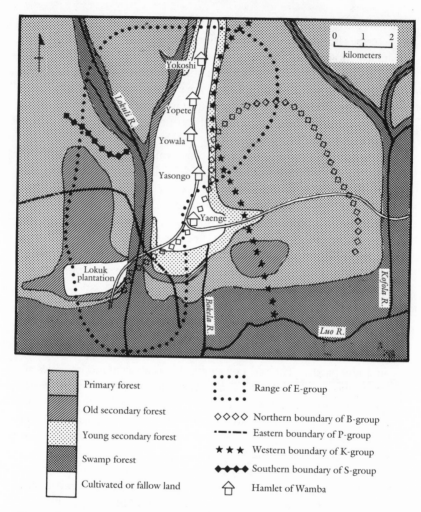

Map 8. Our research site at Wamba.

accompanied by children, providing the origin of the human family. During the provisioning of Wamba's E-group, however, it became clear that my thoughts required revision as individual identification progressed and as we came to understand kin relationships.

From the beginning of provisioning, E-group graphically took the form of a fission-fusion group. At first, (about 1976) there were just 20 chimpanzees, whose range centered around the middle to lower basin of

Table 7. Composition and copulation rate of parties of Wamba's E-Group
(Kano, 1982a)

Party size	Number of parties observed	Average number of individuals of each age-sex class per party				Percent of parties including at least one estrous female	Percent of parties where at least one copulation was recorded
		Sexually mature male	Sexually mature female	Juvenile and infant	Total		
0–5	3	2.0	2.0	1.0	5.0	100	33
6–10	39	3.9	2.2	1.6	7.7	92	36
11–15	16	4.9	5.3	3.6	13.8	100	69
16–20	39	6.8	6.2	4.8	17.8	100	49
21–25	31	8.6	8.6	6.0	23.2	100	68
26–30	27	10.0	10.1	7.4	27.6	100	82
31–35	15	12.0	12.0	8.6	32.6	100	80
39–40	2	12.0	14.0	11.0	37.0	100	100
Total	172						
Mean		7.2	6.8	4.9	18.9	98	59

Table 8. Pygmy chimpanzee age classes
(age above 15 years old is estimated)

Age class		Age (years)
Infant		0–1
Juvenile:	Early	2–4
	Late	5–6
Adolescent:	Early	7–8
	Middle	9–12
	Late	13–14
Adult:	Early	15–19
	Middle	20–30
	Late	31–+

the Lokuli River. Sometimes, however, another, larger group of 30 individuals came southward from the upper basin of the Lokuli and joined with the southern chimpanzees into a very large party. After several days to two weeks, the northern chimpanzees returned to the north leaving the southern ones. Thus, from the beginning of observations E-group was comprised of these two subgroups, which I shall call, for convenience, the northern subgroup and the southern subgroup, respectively.

A juvenile begs for food from an adult while maintaining contact with his mother.

The existence of two subgroups makes the social structure of E-group somewhat complicated. The relationship is probably easier to understand by viewing the respective subgroups as having the character of independent unit groups. That is because the usual acts of fission and fusion are occurring between these respective subgroups. The only difference is that sometimes these two units form one large gathering.

Relative familiarity between individuals always seems to play a large role in the form of a normal party. Although certain individuals always interact with each other, they may have little to do with others. If they do interact with others, they soon separate.

The structural elements of the unit group are divided into male and female sexual classes. As can be seen in Table 8, these are further divided into age classes called adult, adolescent, juvenile, and infant. The highest level of intimacy is between the juvenile and infant offspring and their mother; they always move together, and separate only in exceptional cases. In this way, the mother and the mature son are linked together by continuous strong bonds. From adolescence to adulthood, the male almost always stays close to his mother. A male seen in a small party that has a pairlike structure often is the son of that female. Because the son does not copulate with his mother in this kind of small party, the mixed form is superficial and does not have a reproductive function. From this point of view, my early prediction that the pygmy chimpanzee group form may be the prototype of the human family was incorrect.

The mother-offspring family, which includes the mother and her offspring of both sexes—infants, juveniles, and sons up to and beyond adolescence—becomes very firmly bonded. Because they are seldom separated, it is probably safe to view the mother-offspring family as one fission-fusion unit. In the 1981–1982 records, mother-offspring units accounted for a total of 18 parties, including 50 individuals altogether in the northern and southern subgroups (inside one subgroup, two individuals from P-group temporarily participated). They occupied 67% of the total number of recorded individuals (75) in E-group that year.

Intimate relations are also seen between adult females. The northern subgroup included a middle-aged female who was thought to be sterile and an adult female who joined temporarily and had no offspring. Other than these two females, all the rest were mothers. Consequently, we can probably consider virtually all adult females to be mothers. Therefore, the most important factor determining the party composition may be the intimate relations between mothers.

Males of unknown blood relationship are all in the middle-to-old-age class. We presume that they are probably males whose mothers died. Eight of these males are found in the northern subgroup and four in the

An adult female grooms her adolescent daughter. Her son stands behind her with his arm on her back.

southern subgroup. There is an intimate relationship between them. Three individuals in the southern subgroup spend a great deal of time together, as do five in the northern subgroup. All three from the southern subgroup are often ignored by the females, but when the five from the northern subgroup assemble, they aggregate with many females. These five also appear to be at the group's core. The other males seem to follow a mother-offspring party or other males separately.

Young nulliparous (adolescent) females are opportunistic and their bonding attachment is weak. They may join any party and then enter a different unit group. These females, however, do not form the nucleus of any party.

Thus, in pygmy chimpanzee unit groups, there are three structural arrangements: the mother-offspring unit, adult males of unknown blood relationship, and young nulliparous females. The cohesive level and the level of continuity of the party are apparently decided by the relative degree of intimacy within and between these arrangements. The mother-infant unit, however, may be the most important.

Wamba's E-Group

At the beginning of provisioning, we could not identify all of the individuals of E-group, but we assumed that there were approximately 55 individuals including unsettled adolescent females. Although other groups (P, B, K, and S) had all or part of their group ranges in Wamba forest, we did not recognize as many individuals in those groups as we did in E-group. If we infer their group size from their normal party size, however, we think that P-group was almost the same size as E-group; B-group may have had 80–120 individuals, and K-group, about 100–150 individuals. S-group was rarely seen around Wamba forest, but when it came, it was a large party of 70–80 individuals. S-group may be about the same size as K-group. If these estimates were close to the actual numbers, E-group may have been on the small side among the unit groups of pygmy chimpanzees at Wamba. Subsequently, however, the number of individuals in E-group continued to increase and reached close to 70 individuals at the beginning of 1985. Between 1976 and 1985, 17 infants were born (8 males, 9 females).

We have records about the birth interval in females in only three cases. The intervals were 2.5 years, 4 years, and 4.5 years. Because only three individuals gave birth twice, out of 13 multiparous females observed since the beginning of our investigations, the above data are inadequate as representative values of birth-spacing. More likely, these data are close to the smallest value for birth interval in the pygmy chimpanzees of Wamba. The interval of "2.5 years" is a special case in which the first infant died; the mother of the dead infant clearly ceased to lactate and quickly resumed menstruating.

Almost all females, after being individually recognized, produced offspring or gave birth only once during our research. If we calculate for each multiparous female the interval during which there were no births, starting from the time that each was first individually recognized until the end of the last observation (February 1985), the average value is 3.9 years (in this method of calculation, I add one to the number of births and divide by the number of years of observation of each female). This figure must be too small as a mean value of the average birth interval of females because we started at 0 years, not knowing the actual age of the female's offspring; after that, we calculated by acts of birth. Considering this, our estimate of 5 years as the average birth interval is probably too low.

During the research period from January 1976 to February 1985, there were two confirmed deaths. One was that of a female infant more than 1.5 years old. The other was a male who had just reached adulthood; he was shot by a poacher's gun.

The average annual birth rate and death rate of Wamba's E-group are 3.4 and 0.4 per 100 individuals, respectively. The birth rate is high; the death rate is extremely low. Some people believe that provisioning is one of the reasons for this imbalance. It is known that in Japanese monkey troops where provisioning occurs, the increase in numbers of individuals is phenomenal. The pygmy chimpanzees of Wamba, however, are provisioned for no more than two to three months in an ordinary year. This contrasts with the sightseeing areas of Japan, where monkeys obtain large quantities of gift-food (treats) almost every day. Moreover, in the fruiting season of *Landolphia* (batofe) and *Dialium*, the pygmy chimpanzees disregard the sugarcane that we use in the provisioning process, and forage naturally instead. Thus, the time they strongly depend on artificial feeding is usually about 1.5 months. We do not think that this short period of artificial feeding can completely account for the improved nutritional state of E-group chimpanzees or that it causes the extreme imbalance between the observed birth rate and death rate.

Instead, we think that changes in the relations between chimpanzees and the resident human population are more likely the cause. After we began investigations at Wamba, the villagers stopped injuring and threatening chimpanzees. The chimpanzees also became accustomed to people. Before habituation, when they saw a person, they quickly ran away. Now, in the presence of observers, they will remain and take food leisurely for long periods of time. With decreased pressures from humans, the chimpanzees can also take food when they like at their favorite places in the forest. These are probably the important factors that greatly improved their nutritional status, enhancing overall health and survival.

Emigration, mainly of young nulliparous females, occurred. From 1976 to 1985, at least 30 young females had contact with E-group. Almost all of them disappeared after staying intermittently, for various lengths of time. There may also be, very exceptionally, multiparous females who have contact with other groups. Up to 1985, I have documented at least three females like this who came from another group to E-group. But they all seemed to stray in for short periods only, and they disappeared within a day or two.

During our observations, only five females gave birth for the first time in E-group. The adolescent period can be called a female's wandering stage, but toward the end of adolescence, she will be drawn to some group. Eventually, she seems to intensify her attachment to a fixed group until, after giving birth, she becomes a fairly stable group member. At the beginning of provisioning in E-group, there were seven female juveniles and infants. They grew up, and when they became adolescents, they vanished, one after another. At the end of 1984, only one individual remained. Where they went is unclear, but we think they probably left and entered another group.

Males, unlike females, do not emigrate. The five adolescent males we identified at the beginning of provisioning all grew up in E-group, and they are now adults (except for the one, mentioned above, who was killed by poachers). Additionally, all nine of the male juveniles and infants born in E-group became adolescents in E-group; three of them have passed adolescence and are entering adulthood.

These nine years of transitions within E-group composition are described in Table 9, although during the first two-year term, individual identification was imperfect. An increase in the number of individuals is seen in every sex and age class.

A special characteristic of adults is the almost constant, 1:1 ratio of males to females. By contrast, in the adolescent stage, the number of females of almost every age exceeds the number of males (sex ratio = number of males/total number: range, 0.44–0.31; mean, 0.39). Perhaps this imbalance comes from the unsettled, wandering adolescent females contacting more than one group. Considering this circumstance, we think that the ratio of males and females in the adolescent stage is probably also 1:1.

Recently, it appears that E-group has split or is in the final process of splitting. According to Takeshi Furuichi (pers. comm.), the southern (E_1) subgroup and the northern (E_2) subgroup rarely encountered each other during his research from 1983 to the beginning of 1984. When they occasionally did get together, a large dispute would usually ensue, and soon they would separate. Later, during my investigation (from November 1984 to February 1985), the two subgroups did not contact each other even once.

Up until about 1982, the subgroups had joined regularly, about once per month. We cannot yet deny the possibility that if the northern (E_2) subgroup advances south, it may again unite with the southern (E_1) subgroup as in the past. E_1-E_2 encounters, however, have become increasingly infrequent. Although we cannot say that group fissioning is complete, we can at least say that they are approaching dissolution.

It is noteworthy that almost all the constituent members first recognized in 1976 as belonging to either the northern or southern subgroup are being allocated to the present respective groups without an exchange of members. There are just two exceptions. A female called Lan, who was an adolescent in 1976 and was mainly in the southern (E_1) subgroup, is at present in the northern (E_2) subgroup. Conversely, a female named Bihi, who was clearly an adolescent in 1977, has transferred from the north to the south. Both of them have become mothers of one offspring.

If we decide that the fissioning of E-group is complete or well advanced, the nuclei of the two splitting subgroups must have already existed nine years ago in 1976. Thus, the fissioning of a unit group of pygmy chimpanzees may take a very long time and probably is one example we can offer of slow change. We do not see other examples of this kind of slow group fissioning in other primates.

The composition of the E_1 subgroup in February 1985 is outlined in Table 9. The sex ratio in adults is 0.5, in adolescents, 0.38, and overall, 0.45. The same number of males and females is maintained, except among adolescents, and is a distinctive feature of the pygmy chimpanzee.

Unfortunately, we are unable to check carefully the present membership of the E_2 subgroup because of infrequent encounters with it since 1985. If we infer, however, that until 1984, the whole E-group maintained the same number of males and females and that the same was true in the E_1 subgroup when it separated from E-group, then we expect that E_2 also

Table 9. Number and proportion of individuals by age-sex class in Wamba's E-group

Time period	Adult male	Adult female	Adolescent male	Adolescent female	Juvenile/ infant male	Juvenile/ infant female	Total	Adult sex ratio	Adolescent sex ratio	Sex ratio for all classes
1976–77										
E_1	5	5	2	2	3	2	19	0.5	0.5	0.53
10/77–3/78										
E_1	5	5	2		4	2		0.5		
E_2	9	10	3		5	5		0.47		
Total	14	15	5	8	9	7	58	0.48	0.38	0.48
1/79–2/79										
E_1	5	5	2		4	1		0.5		
E_2	11	10	4		6	5		0.52		
Total	16	15	6	8	10	6	61	0.52	0.43	0.52
1980 (Kitamura)										
E_1	5	5	3		4	2		0.5		
E_2	11	11	4		8	6		0.5		
Total	16	16	7	9	12	8	68	0.5	0.44	0.51

1981–82											
E_1	7	8	2		3	4			0.47		
E_2	13	12	3		7	5			0.52		
Total	20	20	5	11	10	9		75	0.5	0.31	0.47
8/83–2/84 (Furuichi)											
E_1	8	7	2		3	4			0.53		
E_2	13	11	5		5	3			0.54		
Total	21	18	7	11	8	7		72	0.54	0.39	0.5
10/84–3/85											
E_1	7	7	3	5	3	4	*1	30	0.5	0.38	0.45

E_1 = Southern subgroup E_2 = Northern subgroup
Sex ratio = number of males/number of females
* = sex undetermined

retains an equal number of males and females, even after fissioning. In some fissioning of groups of Japanese monkeys, there is an imbalance of the sex ratio on one side of the fissioning troop. But in pygmy chimpanzees, there seems to be no distortion of their social composition.

There is a possibility that, before their habituation, E_1 and E_2 were not constituents of one unit group but were independent unit groups. Because an exchange of members, except for adolescent females, did not occur between the two groups, we cannot discount this possibility. Between 1976 and 1984, frequent sexual and friendly interactions, as well as agonistic ones, were observed during E_1-E_2 encounters. If we assume that E_1 and E_2 were different groups, these interactions have important implications for understanding the transition from pongids to hominids. Among primates, diverse and frequent intergroup interactions are seen only in humans. The intergroup relationships among pygmy chimpanzees may be highly advanced and close to that of human society, but we do not understand enough to settle this issue. New long-term investigations about other groups are necessary.

Environmental Pressures and Physical Abnormalities

From the start of our research, we knew that there were many handicapped individuals in each group at Wamba. One individual lacked a hand; another was missing a foot, with only a stump below the ankle. Such losses of hands and feet were conspicuous and observed from the start. By using binoculars and a telescope to magnify features ($20\times -60\times$), we became aware that the total or partial loss and malformation of digits occurred frequently. There were also many examples of abnormalities of the ears, eyeballs, genitalia, and other parts.

Initially, these physical abnormalities were recorded to aid in individual recognition. Because so many were affected, however, I decided to record the extent and position of each physical abnormality in about 100 individuals (70 individuals of E-group and some from P-group) from November 1981 to February 1982.

Abnormalities of the limbs and digits. The most numerous abnormalities were of the hands, feet, and digits. Total or nearly total loss of hand or foot was seen in four individuals (two males, two females). In one of the males, Masu, a foot was missing, and both legs were paralyzed in a doglegged form. Also, a female of P-group, Mahi, had a leg that was paralyzed and immobile.

A female missing all of the fingers on her right hand.

The following kinds of abnormalities of the digits of the hands and feet were recorded: total loss (28 examples) or partial loss (96 examples), cases in which the affected digits moved at a different angle from the normal ones (nine examples), enlargement (16 examples), reduction (two examples), contracture (nine examples), and permanent dislocation (one example). In addition, brachydactilia was seen in two males of P-group; the second and third toes were half the length of the fourth toe, and they appeared to include only one joint at the base, although they retained normal nails.

Of the 96 individuals examined for the existence of limb defects, 46 individuals had at least one abnormality of the limbs or digits. Differences in frequency were seen between the sexes. More males had defects (63.6%) than females (34.6%) ($P < 0.01$). Of 1,920 digits that were noted, there were 166 abnormal fingers and toes (8.6%) but, again, clear sexual differences in frequency were evident (males, 17.5%; females, 5.5%; $P < 0.001$).

In males, the frequency of defects of the toes was higher than for fingers, but in females a clear difference was not seen. In toes, defects of

A male missing his left foot from the ankle down.

the big toe (hallux) were seen less often than those of the other toes. The same result (fewer defects of thumb than of other fingers) was obtained for the fingers, but the difference was not significant.

In addition, there were differences according to age. The frequency of occurrence, low in infants and juveniles, increased with age (Table 10). After reaching full adulthood, almost 100% of the males (21/22 individuals) had a defect of some digit. In fully adult females, 68% (11/16 individuals) had digital defects.

Many reports suggest that congenital abnormalities of the hands, feet, and limbs occurred and increased in provisioned Japanese monkeys because of exposure to agricultural chemicals. In the pygmy chimpanzee, the only deformity thought to be congenital is brachydactilia. Because the other deformities increase with age, we believe that almost all abnormalities result from environmental influences, the most important of which are treated next.

Table 10. Proportion of individuals affected by limb abnormalities (Kano, 1984b)

Affected group	Infant	Juvenile	Adolescent	Young adult	Middle-aged adult	Old adult
Males						
Number of individuals with abnormalities	3	7	6	6	16	6
Number of individuals missing more than one digit	1	1	2	3	15	6
Proportion (percent)	33.3	14.3	33.3	50.0	93.8	100.0
Females						
Number of individuals with abnormalities	6	6	15	9	9	7
Number of individuals missing more than one digit	0	0	4	3	7	4
Proportion (percent)	0	0	26.7	33.3	77.8	57.1

First are traps, of which snares, especially those made of wire, are most dangerous. If something as big as a pygmy chimpanzee is caught in a snare, it will escape. But something like wire lasts for several months on a hand or foot. When the animal wrenches itself from the sprung tree, the wire bites hard into the skin, and the hand or foot becomes impossible to use. According to African hunters, when that kind of condition continues, the part extending beyond where the wire binds may eventually rot and fall off. Up until now, four individuals have been observed walking with wire attached to their hands or feet. After the wire came off, the limbs and digits did not fall off; nevertheless, three of the individuals could not move the injured part and remained crippled.

Gibbons are called aerial acrobats and are agile inhabitants of the trees, but when their bodies are examined after death, traces of bone fractures are found at a very high rate (Schultz, 1956). Pygmy chimpanzees also spend much time in the trees, and many of their abnormalities, like those of gibbons, may be caused by accidental falls while traversing or foraging in the trees. When fighting, they seek escape in a reckless manner, sometimes miscalculating and falling from heights of more than 10 m. Finger dislocations and bone fractures also probably occur; if the animal bites into the affected part, the end could completely come off.

Some kinds of illness, leprosy for example, may be responsible for some of the losses. Like many other human contagions, leprosy is capable of infecting chimpanzees. Moreover, leprosy is not an unusual affliction in the human population of the Congo forest, including the Wamba area, where it is called *bakechi*. However, many abnormalities of the digits in the pygmy chimpanzee are not progressive. If leprosy is the cause of some lost digits, the proportion probably is not large because of the few symptoms of skin disease.

Perhaps one cause cannot be found for all the abnormalities of the limbs and digits. The frequency of occurrence of abnormalities, however, may be higher in males than in females no matter which cause is assigned responsibility. As I will discuss later, males engage in more aggressive defensive interactions than females. Consequently, accidents during flight and pursuit probably occur more often in males. During group travel on the ground, the ones at the front are males. A male, just before or just after travel, often displays "branch-dragging" in which he holds a branch that was snapped and runs past others for several meters to tens of meters. As a result, males get caught in traps more often than females. Also, the frequency of contracting leprosy is expected to be higher in males, if it is the same as for humans. (In Africa, the sexual differences are small, but in many other regions the prevalence of leprosy is twice as high in human males.) From these points of view, the sexual differences that arise in the abnormalities of the limbs and fingers and toes are probably natural functions of behavior.

Abnormalities of parts other than the limbs. The loss of the testicles was seen in three males of E-group. Because the abnormalities differed in each case, the cause of the loss was also thought to differ. The first one, Masu, did not have a penis or testicles. His nipples were clearly male in form, but there were no external genitalia. As previously mentioned, he was a heavily handicapped individual. Both legs were bent and did not move, and the skin on his belly had become white with keloid. The loss of genitalia and abnormalities of the legs may have been caused by some disease.

The second male, Hata, had a normal penis but no testicles; the entire circumference of his penis was red from inflammation, and flies were always swarming around it. This symptom continued for several years following individual identification but, later, the inflammation disappeared. Hata copulated sometimes. Once, after copulation, I was startled to see blood dripping from his penis. It is unclear whether the blood was from menstrual flow or from his penis. Still, he dragged himself along

An adult male lacking testicles.

with an erect penis from which a great quantity of colorless, transparent mucus came; the sight was bizarre. The loss of his testicles probably was caused by some disease, or they could have been bitten off during a fight.

The third male, Koshi, had part of his testicles concealed by hair but did not show symptoms of skin lesions as in the previous two individuals. He had an underdeveloped penis that was $\frac{1}{5}$ to $\frac{1}{10}$ the size of a normal male's. Perhaps his condition was bilateral undescended testicles (cryptorchidism), a congenital condition in which neither testicle descends into the scrotum. In B-group, one example was also seen of unilateral cryptorchidism.

A variety of abnormalities in parts other than testicles also were observed. One juvenile male of E-group and one adult male of P-group had white pupils that may have been cataracts. The male of P-group clearly had a condition that was progressive.

An adult male of E-group seems to have been afflicted for a long time by a nasal disorder. Pus was seen in his nostrils, and the neighboring skin had turned white. Apparently it was hard for him to breathe through his nose, and he usually had his mouth open.

In P-group, there was a male whose entire face was covered with dark-red scablike things. It was such a horrible disease (name unknown)

A male with a terrible, disfiguring skin affliction.

that I could not look at it squarely; his nose and both eyes were completely ravaged, he had no lips, and his whole mouth was deformed. Furthermore, there has been no change in his condition since individual recognition up to the present (1985) during which several years have passed.

Between 1976 and 1980, a juvenile and an infant had blisters that were prevalent around the lips, the inner part of the limbs, and the buttocks, in a ring-shape. Recently, the blisters seem to have become infrequent.

Several individuals who were past prime had fist-sized tumors on their skin. The cause of these is unclear.

In addition, some individuals have non-disabling abnormalities such as extra nipples or whiteness on part of the skin of the fingers, toes, and face. There are few reported examples of these abnormalities, and they are probably hereditary.

Lacerations of the ear were seen at a higher frequency in males (24/88 ears) than in females (8/102 ears). The main cause for that may be aggressive interactions.

The physical handicaps and the numbers of afflictions just described in the pygmy chimpanzee groups at Wamba are not peculiar solely to Wamba. I observed three individuals without either hands or feet at Yalosidi, and American researchers at Lomako have also seen a female lacking a hand. If research at the detailed level applied at Wamba becomes possible in other regions, we may be able to explain their high frequencies.

It is noteworthy that for practically every possible illness and abnormality, morbidity is higher in males than in females. That is, environmental pressures may be more severe on males than on females. According to the data, however, slightly disabled individuals and even severely crippled individuals manage to survive for a long time. The environmental pressures on the pygmy chimpanzee, at least, are not great enough to have fatal consequences.

Comparison with Common Chimpanzees

The size of unit groups of common chimpanzees at several research sites has been reported (Table II). Some are exceptionally large, such as the M-group at Mahale, but as a rule, the pygmy chimpanzee tends to form larger unit groups than the common chimpanzee. We think that the two species probably differ in the extent of mutual tolerance.

For example, there are differences in party size. In the common chimpanzee, a party generally consists of one to five or six individuals; this is exceptionally small compared to the party size of pygmy chimpanzees at Wamba. There are reports that the pygmy chimpanzees of

Table II. Unit group size, composition, and sex ratio in common chimpanzees

Unit group	Adult male	Adult female	Total	Adult sex ratio	Reference
Gombe					
(Before fissioning of unit group, 1965–1971)	16	12.3	51	0.57	Goodall (1983)
(Kasakela Group, 1972–1980)	7.4	16.3	48.5	0.31	
Mahale					
(K-group, 1962–1973)	5.4	9.3	28.4	0.37	Nishida (1979a);
(M-group)	15	30+	80	0.33	Kawanaka (1979)
Budongo	25	30	85	0.45	Suzuki (1977)
Bossou	6	8	28	0.43	Sugiyama and Koman (1979)

Lomako Forest and the west bank of Lake Tumba form small parties of two to five individuals (Badrian and Badrian, 1984; Horn, 1980). At Lake Tumba, however, the population has been extremely reduced by people. In the Lomako Forest, there has not been enough habituation, and because the accuracy of observation is consequently low in these conditions, we must wait for future data. Judging from the study at Yalosidi and from my own survey, we are unlikely to be mistaken in our view that the pygmy chimpanzee has a more gregarious disposition.

In addition to differences in gregariousness, both species clearly differ in their pattern of group fission and fusion. In the pygmy chimpanzee, what we call a party is essentially limited to the "mixed type," which simultaneously includes a male, a female, and offspring. By contrast, in the common chimpanzee, three forms of party besides the mixed party occur in high frequencies: the "bisexual adult form," which consists of other males and females but does not include mother and offspring; the "mother-infant form," which consists of mothers and their offspring; and the "male form," composed of adult males only. As a consequence, the percentage of "mixed" parties is lower than in pygmy chimpanzees; at Mahale, it is 47.7% of all the parties (Nishida, 1977) and at Gombe, it is 30% (Goodall, 1968). Again, in the common chimpanzee, males and females will frequently temporarily move alone, but in the pygmy chimpanzee, they rarely do so.

Richard W. Wrangham (1979) presented a simplification of the flexible group formation of the common chimpanzee. Here, I shall introduce and summarize his model. A female common chimpanzee is anestrous for several years after giving birth, until the offspring can almost take care of itself. During this time when she does not accept the sexual advances of males, the female accompanies only her offspring and leads a life confined to just her own narrow region. This kind of mother-offspring family divides the habitat into small parcels and survives by being independent of males.

On the other hand, males have a large home range encompassing several home ranges of mother-offspring units, around which the males forage widely. The mother-infant units interact with males when they enter their home ranges, but when the males leave, they do not follow them. As the offspring mature and require less care, however, the females resume estrus. They abandon their narrow mother-infant ranges and go in search of males. They get pregnant, deliver, and when estrus ceases again, they return to their own home ranges.

Males of a unit group are bound by strong ties. They will gather and oppose the males of neighboring unit groups. A female, however, may separately accompany whatever male comes into her range and follow him. A female transfers between unit groups and, consequently, her position appears unstable. The "unit group" (community) is, therefore, a general idea that can be applied only to males in common chimpanzee society; females are basically independent actors. This interesting scheme also suggests continuity between common chimpanzee society and orangutan society where both males and females are independent. This is Wrangham's way of looking at common chimpanzee society.

Pygmy chimpanzee society, which forms only "mixed" parties, appears limited in its freedom when compared with common chimpanzee society. Because of that, pygmy chimpanzee society seems to be a compromise form between common chimpanzee society and gorilla society. However, when we look in detail at the system of formation of pygmy chimpanzee parties, we see that it is different from gorilla society. First, the most important fission-fusion unit of the pygmy chimpanzee is the mother-infant family, but this is also true for the common chimpanzee. A point in which they differ is that in the common chimpanzee, only the mother and her infant and juveniles are included in mother-infant families, but among the pygmy chimpanzees, even mature sons join in.

The gorilla and common chimpanzee have in common males as a second type of fission-fusion unit. In the common chimpanzee, adult males are intimately bonded. This "male bond" is an important keystone supporting common chimpanzee society. The males of one group join together and oppose males of other groups; they will try to attract estrous females, if they are present, and engage in a reproductive session. Because the females are enticed by the males, mutual independent friendship among females is extremely low. The gorilla unit group consists of one adult male and several adult females, who are drawn to the male, and their offspring; females are mutually indifferent and may even be unfriendly. Female common chimpanzees are very similar to gorilla females in disposition.

On the other hand, since young adult male pygmy chimpanzees follow their mothers, the male bond between them is not as firm as that between common chimpanzee males. But friendship between adult female pygmy chimpanzees (that is, mothers) is great. Mother-offspring families gather and usually form a large party. Because the party includes adult sons, large multi-male parties are formed that become natural reproductive units. In this way, the mother is the core of pygmy chimpanzee society, and the males lead a life following their mothers.

] 91 [

Thus, although the pygmy chimpanzee differs completely from the gorilla in that the female pygmy chimpanzee is central in the formation of parties or groups, the common chimpanzee resembles the gorilla in that males form the nucleus of the unit and females are not friendly with each other. Therefore, pygmy chimpanzee society resembles a blend between that of the gorilla and the common chimpanzee, while common chimpanzee society is probably intermediate between the gorilla and the pygmy chimpanzee form.

One more difference between pygmy and common chimpanzees is that in the pygmy chimpanzee, young nulliparous females are practically the only ones who move between unit groups. In the common chimpanzee, both nulliparous and parous females can transfer between groups. Because of this, common chimpanzee society is more fluid and complex.

The greatest difference between the societies of the two chimpanzee species is the nature of individual survival. Among pygmy chimpanzees, the males may experience slightly greater environmental pressures than females, but the survival rate is not different; the male ratio for any age class is right at 0.5. In marked contrast, among common chimpanzees, the adult ratio is 0.31–0.45; that is, there is one male to a number of females, specifically 1.2–2.2 females (Table 11). This ratio shows that among common chimpanzees, males are thinned out in higher proportion than females by various environmental pressures. Several causes of death are clear, but the most important, to be discussed later, is conspecific killing; that is, the killing of one member of a species by another member of the same species.

I will conclude by saying simply that the pygmy chimpanzee has a higher aggregative nature than the common chimpanzee; that is, the pygmy chimpanzee is more gregarious. When the tendency to aggregate is high, we would expect that social frictions and discord must be great as a consequence. Nevertheless, such problems are apparently resolved smoothly in pygmy chimpanzee society because the individual survival rate is very high. In short, the pygmy chimpanzee seems to excel at individual coexistence.

I want to defer to the following chapters the investigation and consideration of such questions as, what kind of structure underlies a society of coexisting individuals and why is gregariousness suited to the pygmy chimpanzee?

CHAPTER FOUR

Food and Nutrition

The Variety of Foods

Primates, in general, are regarded as mainly plant-eaters. Among herbivores, there are monophagous species, such as koalas, who are said to eat only the leaves of eucalyptus. Primates, however, will not tolerate such a monotonous diet. No matter at which primate we look, it probably utilizes a large number of plants as food.

Plants germinate, their leaves grow thick, their flowers bloom, they bear fruit, and they scatter their seeds. When the appropriate season comes, some plants drop their leaves while other plants die back. In this way, plants variously change their guise in response to the respective phenology. This is the same for vegetation of regions that have four seasons, for vegetation when there is a dry and a rainy season, or for vegetation in limited humid tropical regions where climatic change is barely visible.

Primates eat not only many species of plants but various parts of the plant as it changes form in response to the season. By so doing, primates were able to adapt well to the rich complexity of the plant world (especially the forest) where "diversity" is the special feature of their food.

Plants impart a completely different taste depending on the species and the part eaten, as do different meats. It has been said that primates are animals with the most developed sense of taste. That is probably because it was necessary to identify the flavors of various species of food.

The food list of pygmy chimpanzees recorded at Wamba consists of 147 items (Appendixes 1 and 2) and the number will probably continue to increase as we gather more data. Although pygmy chimpanzees subsist primarily on plant food, including the flesh of fruit, seeds, sprouts, leaves, flowers, bark, stems, pith, roots, and mushrooms, they also eat small mammals (flying squirrels, etc.), insect larvae, earthworms, honey, eggs,

A pygmy chimpanzee feeding on *Dialium*.

and soil. Fruits (pulp, seeds) occupy 57% of the whole list and probably form the core of their diet when considering only food type.

A special problem that continues to bother researchers in primate ecology, however, has been how to quantify actual intake. Up to the present, the methods traditionally used to investigate intake are (1) estimation according to the duration of feeding, (2) estimation using the quantity of the contents of their feces, and (3) estimation by use of both sets of information. Each method has perplexing shortcomings.

The quantity of food that is taken in during a set feeding interval— intake efficiency—differs greatly depending on the degree of enthusiasm or hunger. Also, even if the enthusiasm is the same, the intake efficiency may differ depending on the nature of the food. As an example, compare how someone eats a large football-size fruit with plenty of flesh attached to how one eats small fruits having skin that must be removed diligently, piece by small piece.

Differences in observation conditions are also important. For example, when chimpanzees are eating high in a large tree, frequently several individuals are in the field of vision simultaneously for a long time. When they are feeding on the ground in dense forest undergrowth, however, occasionally one individual will be visible and then disappear. Thus, we cannot know exactly the absolute quantity, nor the relative quantity, of food consumed from just the apparent feeding duration.

By studying the contents of feces, we can compensate for some of the weaknesses of direct observation since the results of feeding activity over a long time are compressed. This kind of analysis is not affected by differences in observation conditions, but there are defects. Seeds that have been eaten whole together with their flesh can be readily detected in the feces. But fecal analysis is exceedingly difficult to undertake when seeds or things crunched by teeth and eaten (for example, beans), stems/stalks, leaves, pith, and fibrous parts of the plant body are eliminated as fragments of indigestible parts. In all instances it is certainly difficult to estimate the quantity of plants eaten, even if we can identify what was eaten. Seeds that are too large and that may be discarded after consuming the flesh will not be detected by fecal analysis.

In conclusion, to figure accurately the amount of food consumed by a species, it may be necessary to follow one individual all day, and to do nothing else but diligently record how much and what it eats. In order to do that, the focal individual must always be kept within the field of view. In the dense forest of present-day Wamba, this is only a hope.

The method I finally used to estimate feeding proportion (the percentage occupied by each food item relative to the whole quantity of food consumed) was a rough compromise. When I discovered an individual who was eating food A, I scored one point for that food. If two individuals were simultaneously eating that food, two points were scored. The scoring occurs from the beginning of observations and continues every time there is a lapse of at least 30 minutes. For example, at 10:00 A.M., one individual is feeding on A (one point); at 10:34 A.M., two individuals are feeding on A (two points); and at 11:20 A.M., three are feeding on A (three points), providing a total of 6 points scored for A. If, on the way, the focal animal samples a different food B, I take that as a new starting point. For example, if an individual who has eaten A eats B and then returns to A within 30 minutes, I will score two points for A and one point for B.

Fecal analysis is handled independently. Feces that are found are wrapped separately and carried back to the research base. After measuring the weight of just the fecal calyx (measurable up to 200 g), we put it in a sieve and wash it in water. We wash away the pasty part; spread what remains on top of a paper; and separate similar kinds of food residues

Table 12. Feeding proportions (percent) (Kano and Mulavwa, 1984)

Food	Direct observation					Fecal analysis, 11/1/81– 2/28/82
	11/1/75– 2/10/76	10/9/76– 12/31/76	10/20/77– 2/10/78	10/21/81– 2/28/82	Mean	
Fruits						
Arboreal	84.3	90.7	79.6	79.1	83.4	86.8
Ground	0	0	0	0	0	6.3
Total	84.3	90.7	79.6	79.1	83.4	93.1
Leaves						
Arboreal	13.4	9.1	15.9	11.8	12.6	3.7
Ground	2.3	0.2	1.7	6.0	2.6	2.0
Total	15.7	9.3	17.6	17.8	15.2	5.7
Animal foods	—	—	2.8	3.1	1.5	1.1

(seeds are abundant), concentrating them into piles. We estimate the quantitative ratio by eye.

The results of direct observations and fecal analyses are displayed in Table 12. As mentioned previously, differences based on the special characteristics of the two methods are apparent, but the obtained ratio of food intake by pygmy chimpanzees at Wamba is as follows: fruits, including seeds, 80–90%; fibrous foods or leaves, 10–20%; other (animal material, etc.), several percent. Thus, when we consider the species of food and the quantity consumed, the most important food is fruits.

Principal Foods and Seasonal Changes

Although we could say that a special feature of the chimpanzee diet is its diversity, when we look at feeding time, we find that they are not taking in many different foods. The average number of food items recorded per day by direct observation is 2.0 types (range, 0–9 types; *N* [number of observation days] = 302 days) and by fecal analysis, it is 6.0 types (range, 1–12 types; *N* = 140 days).

The number of food types that are recorded by direct observation increases with the length of observation time. In fecal analysis, when the number of examined feces reaches 5 or 6 samples, the number of food types does not increase beyond that. In other words, the examination of feces may be a method that surpasses direct observations to assess diet quickly.

When we investigate the number of kinds of foods per month, we find similar tendencies when using the two above-mentioned methods. The number of food types recorded by direct observation is 7–29 types per month (range of time of direct observations, 4.1–108.7 hours; number of months = 18). As the observation time increases, we obtain a curve that increases gradually (Fig. 9). The extreme value of this upwardly curved line is probably an approximation of the actual number of species of foods consumed per month. From Fig. 9, we estimate that it is probably about 40 species.

On the other hand, from fecal analysis, we get 22–27 species per month (range of number of feces, 192–328 samples; number of months = 4). Variation depends on the number of months and the number of fecal samples, but is slight. Consequently, we decided that 200 individual scats was a sufficient sample from which to obtain the upper limit for the number of food species per month as determined by fecal analysis. Because of the defects of fecal analysis that make detection difficult for some foods, this upper limit must be lower than the true value.

Although many foods are included in the diet, this does not mean

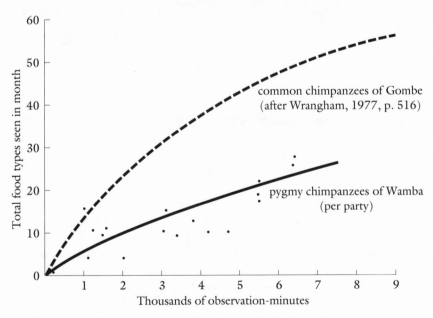

Figure 9. Number of food types eaten per month (Redrawn from Kano and Mulavwa, 1984).

Wamba villagers holding clusters of *batofe* (*Landolphia owariensis*).

One of the trackers, Nkoi, with *ntende* fruit (*Pancovia laurentii*).

that each of these foods has equal importance. If you look at a one-month unit, one species of food alone may occupy an average of 50% of the total ratio of food taken. In addition, the ratio of consumption of the top-ranking six food items, taken together, amounts to 90% of all food stuffs. Pygmy chimpanzees rely on and survive on a small number of food items of what should be called "staple foods."

In general, fruits ripen during a certain fruiting season, but these seasons differ in length. Therefore, the food items that serve as staples fluctuate according to the fruiting season. Some tropical fruits differ from Japan's seasonal fruits by lacking a clear annual fruiting cycle. More-over, various tree species have independent phenological cycles that com-plicate matters. For example, the fruit of a woody vine of the Apocyna-ceae family called *batofe* (*Landolphia owariensis*) normally begins to ripen in about August and becomes a staple food of pygmy chimpanzees at that time; but at the end of September, the fruit disappears. Bumper crops of *batofe* occur in irregular cycles at an interval of from one to three years. In those years, the fruiting season lasts until the end of December.

The majority of tree species that have a vegetation phenology like that of *batofe* bear fruit in large quantities once every several years, according to observations in the Zaire Forest. Because many of the species, like *batofe*, are not abundant, they are significant as food only once every several years. In the Wamba forest, there are at least four species of *Dialium* (locally called *keke, laka, elimilimi,* and *boleka*); they have a stable annual cycle of fruit production between October and February. Every year *Dialium* spp. as a group provide the pygmy chimpanzees of Wamba with very reliable staple foods. By contrast, at Yalosidi, which is 220 km south of Wamba, the annual production of *Dialium* spp. (locally called *lokake* and *elimilimi*) appears to fluctuate a great deal. In November 1973, when I first visited there, *Dialium* fruit was the predominant food of the chimpanzees. According to what I heard later, that fruiting period lasted until about March of the next year. However, during my subsequent research from November 1974 to February 1975, not one morsel of *keke* fruit was seen. Pygmy chimpanzees must survive while faced with this kind of unstable fruit production, and consequently, their food repertoire must be large.

I divided the foods of the pygmy chimpanzee into three grades in accordance with their level of importance, and each month was divided into three 10-day periods. Staple foods were those that constituted more than 30% of the intake ratio in at least one period; semi-staple foods were those that accounted for 30–10%; and supplementary foods, those that supplied less than 10%. In a total of 45 periods of research, the foods that became semi-staple or staple foods, at least once, consisted of an overwhelmingly great number of fruits (23 species), the leaves, stems, and pith of eight species of fibrous foods, and two species of animal food products.

Almost all of the arboreal species that have become principal foods are large trees or woody vines with fruits. For example, trees of the genus *Dialium* are emergents (large trees whose crowns jut out above the line of the tropical rain forest's uppermost stratum). In the peak fruit-bearing period, a large group of 40 pygmy chimpanzees may visit the same tree several times. The woody vines of *Landolphia* spp. are similar to *Dialium* spp. in their pattern of fruit production. The vines climb up around large forest trees and produce numerous fruits the size of oranges on the crown surfaces. Thus, the principal foods of pygmy chimpanzees have the special feature of being produced in large quantities in one food patch. If this were not so, they would not be staple foods.

Among fibrous food items, the most important food types are the shoots of herbs of the family Marantaceae. At Wamba, there are more than ten species of Marantaceae, and all of them send up a slender sprout. Chimpanzees strip off the hard outer husk and eat the tender, white leaf-roll inside. Shoots of *Megaphrynium macrostachyum* and *Haumania lie-brechtsiana* are the most preferred. The shoots of the former are called *beiya* or *beiye* and are also consumed by the villagers, who consider them excellent vegetables. *Beiye* is the main objective of gathering forays that the women of Wamba make every several days. Collected in large quantities and brought back to the villages in shoulder crates, *beiye* can be eaten raw, but if boiled with fish and meat, it has an indescribably delicious flavor; the crunchiness is also good. If we compare it to the vegetables of any other country, I believe it would be considered first-rate. That there is no *beiya* in Japan is truly regrettable.

The shoots of each species of Marantaceae including *beiya* are not seasonal and are frequently fed on by chimpanzees. Because the shoots are scattered uniformly over a wide area, however, they are inconspicuous when fruit foods are abundant. The fibrous foods become prominent as principal foods during the dips between the fruiting seasons of staple fruits.

The larvae of a hesperiid butterfly called *tohilihili* is one of the pygmy chimpanzee's favorite animal foods. These green caterpillars, about half the size of a silkworm, infest the *botuna* (*Cynometra hankei*) trees starting in the middle of December, after a continuous period of scant rain lasting up to one and a half months. During this period when you walk in the forest, you can hear a sound like that of drizzling rain striking a roof, but it is the sound of the caterpillars' fine droppings falling from the large *botuna* trees and hitting the leaves of the undergrowth. Pygmy chimpanzees voraciously eat caterpillars during this period, but the *tohilihili* season lasts, at the most, from only one week to ten days.

Another animal food item is the earthworm. Two species of earthworm are eaten: one lives in the forest floor of *bombongo* (*Gilbertiodendron dewevrei*) groves and the other lives in swamps. The chimpanzees dig and search for them in the dirt or mud using only their hands. In swamp conditions, they dig diligently for earthworms, sometimes foraging slowly for as much as three to four hours. Because this kind of feeding is slow, the amount of time spent feeding is high, but the intake efficiency is very low. In my observations, a male chimpanzee who was carrying something in his mouth took, on average, one every 25 minutes in the swamps and one no more than every 2.5 minutes in the *bombongo* grove. Further-

Caterpillars or *tohilihili* (the larvae of a hesperiid butterfly), eaten by pygmy chimpanzees and villagers.

Millipedes were reported to be eaten by pygmy chimpanzees at the American field site, Lomako.

more, chimpanzees hardly chew when eating earthworms; the shape of the earthworm remains intact in the feces. We think that earthworms probably do not contribute much nutrition at all.

The reason for this kind of unproductive enterprise is difficult to understand. Earthworm-digging is performed enthusiastically, even when there are other foods of the *batofe* fruiting season in abundance. The pygmy chimpanzees do not search out earthworms because they are being driven to eat them from hunger. Instead, they are undoubtedly full enough and, in fact, eat earthworms when there is leisure time. The activity resembles a household of people who leave for the coast on an occasional holiday, amusing themselves gathering shells at low tide and happily returning with a cupful of short-necked clams. In other words, it may be recreation. When you see the figure of a pygmy chimpanzee absorbed digging in the dirt, that is what you may think. It is hard to accept that the merit of this habit is the slippery sensation when the earthworm passes down the throat.

Recently, there was a television report that in the United States, earthworms raised as fishing bait are becoming a popular item for people to eat. Adults or even children, at mealtime, merrily picked out wiggling earthworms, heaped in a bowl, and put them in their mouth. I was dumbfounded, but I was reminded of the earthworm-digging of pygmy chimpanzees. Earthworms may prove to be unexpectedly good to eat.

Taste

The fruits of *Dialium* normally begin to ripen about October, and at this time, the pygmy chimpanzees may start to eat them. As long as the *batofe* fruits last, however, *Dialium* is not consumed much. Consequently, once, after several years when there was an abundance of *batofe*, *Dialium*-eating was delayed until December. When the *Dialium* season passes, the season of *ntende* (*Pancovia laurentii*) begins, extending from the end of January to the beginning of February. Usually, *ntende* begins to ripen shortly after *Dialium* is finished, but in the year when there was a delay because of the abundance of *batofe*, *ntende* began to ripen during the fruiting season of *Dialium*. At that time, the pygmy chimpanzees left the *Dialium* to feed on *ntende*. Then when *ntende* was finished, they returned to the *Dialium* that was left.

In an average year, *Dialium* is an important food that is the most stable of the staple foods of pygmy chimpanzees. Its relative value decreases, however, when the fruits of *batofe* and *ntende* can be consumed. Clearly in chimpanzees, as in humans, taste preference exists.

We think several factors determine food preferences. First, the flavor of the food is important, as it is in the three previously mentioned species of principal foods. But among the foods that the pygmy chimpanzee enjoys eating, many are sweet and sour. The strong inclination for sweet flavors is assumed by humans.

We think that the relative difficulty of the feeding process, the intake efficiency, and related matters also influence preferences. *Dialium* fruit has a stronger sweet taste than the fruit of *batofe* and *ntende*, and it has a refreshing flavor. There is no reason for human judgment to be the same as the chimpanzee's, but when I asked people to make a taste comparison, they said *Dialium* was the most delicious. Compared with *batofe* and *ntende*, *Dialium* fruit has one disadvantage as a good food. Eating it is a nuisance; it is small and the edible part is concealed by a brittle and fragile skin. The work of processing the fruit is tiresome and time consuming. Accordingly, intake efficiency must be low.

There may be other reasons for *ntende* being preferred over *Dialium*. The *ntende* trees in the forest are filled with enormous orange clusters of fruit, which ripen quickly in one to two days. However, the season also ends quickly. Within ten days to two weeks, the fruits completely vanish. As expected, the appetites of chimpanzees and monkeys do not diminish at this speed. At the end of the fruiting season, many *ntende* fruits fall to the ground, and although satisfying in their sweet and sour fragrance, some are left to rot. Eating *ntende* is a race with time.

Dialium, on the other hand, is a long-lasting fruit, with a fruiting season that lasts several months. Some fruits fall to the forest floor, and the outer covering becomes black. But even if they grow moldy, the interior seems to stay edible. I was amazed to see pygmy chimpanzees carefully push aside fallen leaves and select *Dialium* fruit in this condition, one by one, as we might search for mushrooms. With the covering removed, the flesh within certainly does not appear much changed from when it was attached to the branch of the tree. I have not, however, had the nerve to sample the food. Owing to the length of the fruiting season of *Dialium*, it seems like an exaggeration to say that the pygmy chimpanzee exhibits food selection, but it is possible.

Preferences toward foods change even within one day. After arising in the morning, pygmy chimpanzees have fruit as their primary food objective. Then from noon, the sprouts of herbs (Marantaceae), tree leaves, and other fibrous foods are consumed in large amounts. The pattern of consumption in which fibrous foods are used relatively late in the day has been observed also in the common chimpanzee.

This pattern is explained by the existence of poisonous substances, called toxic secondary compounds, in the body of the plants. Plants, which are immobile and suffer from the free attacks of mobile animals, have produced toxic secondary compounds to defend themselves. Because plants must defend themselves throughout the year, especially in the tropical rain forest where seasonality is limited, most plants apparently have these poisonous secondary compounds. The toxic secondary compounds are not found in nectar and pollen. The reason for their absence from pollen is unclear, but nectar was adapted originally for being eaten by animals. Fruit, another part that is meant to be eaten, also contains toxic secondary compounds; there is a need to single out the animals that are seed dispersers. Naturally, there are more toxic secondary compounds included in other parts of the plant body (Janzen, 1975).

Chimpanzees prefer young leaves and sprouts as fibrous food. Among a variety of poisonous secondary compounds, the representatives included in young leaves and sprouts are alkaloids. The quantity of alkaloid within plant bodies has its highest value in the early morning, and declines during the day. Following its daily cyclic rhythm, toxicity starts to accumulate again in the evening. Chimpanzees select foods by adapting to this physiological rhythm of plants, inferring it by some mysterious means.

Foods must be considered from two standards—availability and preferences (we cannot deny that chimpanzees have food preferences). Foods that are found in great quantity and are highly preferred often become staple foods. The feeding habits of chimpanzees may be sustained principally by these foods. Foods that are quantitatively scarce but highly preferred are potentially important foods. If environmental changes favor a large yield of them, they might be adopted by pygmy chimpanzees as staple foods.

The majority of the foods, reaching 147 species in the pygmy chimpanzee, are not highly preferred and are thought to be rather unimportant. Up to the present, most of them have shown up only infrequently in the growing body of research reports. Moreover, all of the foods that will be added to the food list in the future will certainly be "snack"-like items. Snacks, however, are not limited only to a supplementary role in the diet of chimpanzees. They may also be a reserve for coping with repeated environmental fluctuations. In the distant future, some hundred or thousand years from now, the chimpanzee's habitat may be quite different from what it is today. The greater their food repertoire is, the greater will be their chances of survival.

Food Culture

Previously, I emphasized the uniform nature of the composition of the Zaire Forest. Reflecting this uniformity in vegetation, many major foods are common over a wide range. At Yalosidi, the fruit of *bondora* (*Trichosypha ferruginea*) and *bofambu* (*Gambeya lacourtiana*) are often eaten (see Appendix 3 for complete list of plant foods at Yalosidi). At Lomako also, the fruit of *bofambu* seems to be an important food. At Wamba, however, neither is eaten. These regional differences reflect the availability of foods. That is, in the forests of Wamba, these tree types do not exist or they are exceedingly scarce.

Among wild plants, the emergent plant *Ranalisma humile*, called *ikora-yokuwa* at Yalosidi, is an example of an item that is used or not used depending on the region, even though it exists everywhere. The Zaire Basin does not produce rock salt, and long ago, people boiled down this grass and obtained *engange*, a salt substitute. *Ikora-yokuwa* occurs abundantly in the swamps of Yalosidi, and pygmy chimpanzees regularly visit the swamps and eat this grass during periodic fruit shortages. At Wamba, the same grass grows thick on the mud floor of the relatively shallow, wide stream of Bokela. The pygmy chimpanzees at Wamba, however, go there to search for worms; they do not touch the grass.

Regional differences are more apparent in animal foods than in plant foods. At Lake Tumba (Horn, 1980), Lomako (Badrian, Badrian, and Susman, 1981), and probably Yalosidi, termites are on the pygmy chimpanzees' menu, but at Wamba, there is no evidence of this whatsoever. In the forest, one can often see a large millipede, the thickness of a finger and 20 cm long, crawling silently. People of the Mongo tribe call it *kongori* and abhor it as a taboo, a messenger of Satan. However, Badrian, Badrian, and Susman (1981) report that the pygmy chimpanzees of Lomako ate these millipedes frequently at a certain time in 1979. At Wamba, this has not been seen. In 1981, when I met Noel Badrian in Boende and asked him directly, he answered that the Lomako chimpanzees ate the millipedes only in that year. Since then, they had stopped eating them, even though large millipedes are available every year and every season. I wonder if pygmy chimpanzees also have fads in food.

In general, pygmy chimpanzees at Lomako consume more varied

animal foods than at Wamba. At Lomako, snake and other vertebrate animal bones and insect legs and wings often are eliminated in their feces. Recently, N. Badrian and R. K. Malenky (1984) reported that a blue duiker was caught and eaten. As far as mammals are concerned at Wamba, only the bones and a piece of fur of a flying squirrel were found several times in feces. Earthworms, however, which have become relatively ordinary food at Wamba, are not reported as food at Lomako.

The repertoire of animal foods is narrower than that of plant foods. This narrow breadth and the local discrepancies in the animal food repertoire indicate to me that the consumption of animal foods occurred later in the history of the pygmy chimpanzee line. Although regional differences in food habits, called "food culture," certainly exist, clarification must await further progress in research both within and outside the region of Wamba.

Even among the five unit groups residing in the forests of Wamba, we see differences with respect to several foods that are limited by conditions of availability. For example, a principal research focus at Wamba in the early days was B-group. Suehisa Kuroda planted a sugarcane field in B-group's range, but it was completely disregarded by the chimpanzees. In the end, it was raided by elephants. A sugarcane field was also planted in E-group's range, however, and after the sugarcane ripened, the chimpanzees of E-group settled in the field and did not leave until they had consumed all of it. The reason for this difference is clear. Sugarcane has been cultivated in this general region for a fairly long time, but only on a small scale—a little bit in people's backyards, yielding an insignificant amount. Nevertheless, the home range of E-group extends over both sides of the road leading from Yaenge hamlet to Yokosi hamlet, and along that road are intermittent rows of houses. Thus, E-group chimpanzees have many opportunities to encounter the edges of sugarcane fields when they cross the road. On the contrary, B-group's home range extends toward the east from Yaenge and also enters between the road, but on this road there are no houses and consequently no sugarcane fields.

Pineapple presents the same kind of situation as sugarcane. Pineapple is also a foreign cultigen, but it so excels in propagative properties that feral pineapple has gone rampant and spread out into the secondary forest of fallow land. Consequently, pygmy chimpanzees of E-group and of B-group take pineapples as food.

Apparently, pineapple is a more desirable food than sugarcane to the members of E-group, because when sugarcane and pineapple are served simultaneously, pineapple is preferred. When P-group came to feed, however, an interesting thing happened. P-group was divided into two kinds of individuals: ones who dashed toward the pineapples, carrying them away in both hands and fleeing as if afraid of having plundered them, and ones who, quite unconcerned, left with only sugarcane. The individuals who took pineapples were all young parous or nulliparous females, and thus were individuals thought to have recently transferred into P-group. The home range of P-group overlaps broadly with that of E-group, but the former does not extend into the region of the left bank of the Lokuli River where pineapples grow wild. Therefore, the principal members of P-group are unfamiliar with pineapple. By contrast, young females, when they were juveniles in another group or during their adolescent wandering period, had sufficient possibility of coming into contact with pineapple.

Pineapple and various foods other than sugarcane were provided at the feeding site. Remains of small animals and meat did not attract the interest of pygmy chimpanzees at all. Similarly, a live blue duiker and chicken evoked no reaction, except perhaps fear or a threat. Many cultivated fruits were also completely ignored. Only chicken eggs elicited a response resembling, to some degree, that to pineapples. Of 50 individuals who encountered the eggs, 12 carried away one or two. Among those 12, at least 4 individuals bit into, broke open, and ate an egg; the other 38 individuals completely disregarded the eggs. Because the individuals who showed interest in eggs included a variety of age-sex classes such as adult males, middle-aged to old females, and young females, this situation differs from that involving pineapples in P-group. In the case of the eggs in E-group, behavioral differences suggest individual differences in food preferences but not differences between groups.

The females that did not take pineapples, those females past middle-age in P-group, present a problem in understanding the dietary habits of the pygmy chimpanzee. Those females must have been born and raised in a group other than P-group, and some of them must have eaten pineapple while they were young in that group. Those females, who were in their middle years while living in P-group, did not seem to remember pineapples, even though they carried the flavor of home. Because the feeding habits of pygmy chimpanzees are based on this kind of shaky memory, an item may remain in the memory as "food" only if an individual is re-exposed to it within an appropriate interval. In the event that the environment changes, the old food may be forgotten. At such a time, having a wide "range of diet" may provide an important basis for survival. The ability to exploit new, unfamiliar foods may also be important.

The pygmy chimpanzee's tolerance of new food resources may be assessed by how they have accepted newly introduced cultigens (especially fruit-bearing ones). On the basis of my extensive survey, in the regions where pygmy chimpanzees have access to sugarcane, they do eat it. There is, however, not much information from regions other than Wamba about pineapple. Banana, papaya, mango, avocado, sweet oranges, and several other species of fruits are widely cultivated in the Zaire Basin, but there is no evidence that pygmy chimpanzees eat them. Papayas were offered repeatedly to the E-group chimpanzees at the feeding site, but they were ignored, except on one occasion when a young female bit and discarded one after carrying it a short distance. Bananas, which also were offered several times, were completely ignored by all adults; only juveniles approached and sniffed them, without sampling, and only on a few occasions.

The fruit of oil palms in the Zaire Forest is probably one of the easiest cultivated foods for chimpanzees to obtain. Since the early 1920's, when oil palms were introduced into the Zaire basin, they have rapidly spread and now predominate in many secondary forests. Although oil palm fruit is one of the most preferred foods of cercopithecoid monkeys (mangabeys and guenons) in the Zaire basin, there is no evidence, at Wamba or other regions, that pygmy chimpanzees eat oil palm fruit. For whatever reason, pygmy chimpanzees seem to be conservative in exploiting new food resources.

Nutritional Value

Vitamin C. In general, vertebrates more advanced than amphibians can synthesize vitamin C (ascorbic acid). In amphibians and reptiles, vitamin C is made in the kidneys, but in mammals, production has been transferred to a much larger organ, the liver. During the age of prosimians, however, a mutation occurred in primates, that deleted the enzyme L-gulono-oxidase, which is necessary for vitamin C synthesis (Scrimshaw and Young, 1976). Of mammals other than primates, only bats and guinea pigs are known to lack the ability to synthesize vitamin C.

Unable to synthesize vitamin C, primates must acquire it through food. In the absence of suitable food, a deficiency of vitamin C causes the disease called scurvy. Although chimpanzees and monkeys are known to develop scurvy, the artificial diets of captive breeding programs are the cause. In conditions where wild food is normally taken, the development of scurvy seems impossible. G. H. Bourne (1949, 1971) of the Yerkes Primate Research Center estimates that the intake of vitamin C by wild chimpanzees reaches 4,500 mg/day. In humans, who are larger-bodied than chimpanzees, 20 mg/day is sufficient to resist scurvy.

Recently, researchers have hotly debated the medicinal effects of vitamin C use. The effect of vitamin C on the body is not merely a passive prevention of scurvy, but instead it has a broad connection with the maintenance and promotion of good health. As a result, a far larger quantity of vitamin C than the current intake standard may be needed for good health (the current standard differs depending on the country, 20–75 mg/day). The basis of this view is the supposition that animals who can synthesize vitamin C, and certainly primates who cannot synthesize it, consume and use 1 g of vitamin C per day.

The above argument was examined for wild pygmy chimpanzees, who are frugivores, and two predictions were made. First, the fruits and leaves eaten by pygmy chimpanzees will have higher concentrations of vitamin C than those not eaten; that is, the pygmy chimpanzee gathers vitamin C very efficiently. Second, the total amount of vitamin C available in consumed natural foods will far exceed the minimum requirement, and will amount to the several grams per day previously suggested by Bourne.

Because vitamin C is easily destroyed, the analysis of sample material must be limited to fresh items, preferably analyzed on the spot. I approached Ryu Asato, of the Laboratory of Nutritional Studies of Ryukyus University, who agreed to participate in an on-the-spot investigation. With a large quantity of his test equipment and reagents, we rode into Wamba to conduct the first on-the-spot analysis of vitamin C in primate field research.

Batofe fruit was the first to be analyzed. At the time, it was a very important food item showing an intake ratio of close to 100%. The results of the test, however, showed unexpectedly that the vitamin C content was *zero*. Dissatisfied, I asked about the analysis, and Asato said he had used fruits provided by villagers four or five days before, when test preparations were not yet ready. I protested that there was a problem in the samples, and then instructed some villagers to climb trees and gather fresh fruits. This time, vitamin C was detected, but at only 1.2 mg/100 g of edible part, which is a low level, 1/20 to 1/60 that of sweet oranges and 1/65 that of papaya (oranges and papayas were analyzed at about the same time). In the end, however, this was the highest value obtained for *batofe*. Results of later tests, repeated several times, all remained in a range of 0.5–1.0 mg (average, 0.7 mg) per 100 g of edible fruit.

When the *batofe* trees stopped producing, *Dialium* (*elimilimi* and *keke*), followed by *ntende*, became the most important food items. They showed a higher vitamin C content than *batofe*, with an approximate average of 3.7 mg % (3.7 mg/100 g) and 4.7 mg %, respectively; but even so, compared with other fruit, it was never high.

In addition to the above-mentioned important fruits, about 50 kinds of pygmy chimpanzee foods have been analyzed for vitamin C. Among fruits, only *bolengalenga* (a fruit of *Cissus* spp., woody vines) showed an exceptionally high value of 30 mg %. Others generally do not reach 10 mg %, and as a rule, vitamin C content was low compared to that of human cultivated foods. Thus, the first prediction, that pygmy chimpanzees would take foods high in vitamin C, was not borne out.

The second prediction was that the total quantitative intake of vitamin C would be high. In order to determine the amount of vitamin C taken in, we must first determine the respective quantities of the different foods consumed. Because this is impossible at the present stage of research, we must use indirect methods to obtain approximate values.

A starting point is to calculate the basal metabolic rate, which for animals is given in the formula $BMR = 70 \times W^{0.075}$ kcal/day, where W is the weight (kg) of the animal. No wild pygmy chimpanzee has been weighed, however, so we shall use 35 kg, which is roughly 85% of 40 kg, the highest value for an adult common chimpanzee at Gombe Park. If this 35 kg value is used as the maximum weight of a wild pygmy chimpanzee, the BMR comes to 1007 kcal/day.

Animals bred in the zoo do best if they consume calories at twice the BMR, and many researchers suggest a value of three times BMR for animals in the wild. If we take the latter as our model and use 35 kg as the weight of the pygmy chimpanzees at Wamba, we calculate that 3021 kcal/day of energy are being consumed. On the basis of A. G. Goodall's (1977) nutritional analysis of wild gorillas, 100 g dry weight of a sample portion of wild fruit yields an average of 540 kcal (gross calories). Because indigestion and other losses average 46%, the actual rate of ingestion is 289 kcal per 100 g dry weight of fruit. Thus, when only wild foods are used for food, the pygmy chimpanzees of Wamba need 1045 g/day in dry weight.

Using 1045 g/day as a base value, when only *batofe* is eaten, the intake of vitamin C per day is 25 mg; when *Dialium*, 58 mg; and when *ntende*, 309 mg. Thus, vitamin C intake from these major foods is expected to be much less than the 4,500 mg estimated by Bourne (1949, 1971). Only when pygmy chimpanzees eat the fruit of *bolengalenga*, which has an exceptionally high vitamin C content, does the intake per day barely reach 1,200 mg.

Certainly, the calculated value of vitamin C intake depends on the food species, the ingestion (absorption) efficiency, the body size of the animal, and the multiple of the BMR used to calculate the number of calories needed per day under natural conditions. However, wild pygmy chimpanzees are unlikely to consume 10 or 20 times their BMR and, therefore, the amount of vitamin C ingested by a pygmy chimpanzee is likely to be limited to several tens to several hundreds of milligrams per day when feeding on its usual natural foods. The number of days that a pygmy chimpanzee ingests vitamin C at the gram level, if any, is very small, perhaps only several days per year. On the basis of our study of the pygmy chimpanzees at Wamba, we do not agree with Bourne's estimate that more than 4 g per day are normally ingested.

Vitamin C reaches saturation at 1–1.7 mg/kg of body weight. The level of vitamin C intake in pygmy chimpanzees may be sufficient to reach saturation, but does not greatly exceed that level, presumably because there is no need. The defect in vitamin C synthesis that occurred at the level of prosimians was not a grave mutation, because the resulting deficiency could be easily cured by regularly ingesting fresh plant food. Special food habits that selected for foods high in vitamin C were not needed, thus allowing the lineage of primates that had a genetic vitamin C deficiency to achieve the colorful radiation seen today.

Protein. Protein often is a problem in the nutrition of primates, which are plant eaters. On the basis of Asato's analysis, calculating protein content by the same method used to evaluate vitamin C content, the estimated amount of crude protein was 29, 49, and 59 g/day, for the consumption of *batofe*, *Dialium*, and *ntende*, respectively. Because the amount of protein necessary for good health is 1 g/day/kg body weight, protein intake was above the required level for each of the three staple foods.

In general, such plant parts as leaves, pith, and stems have a higher protein content than the flesh of fruits (Table 13). As a consequence, the fiber-eating gorilla has an easier time acquiring protein than does the frugivorous pygmy chimpanzee. According to A. G. Goodall (1977), the Kahuzi gorillas take, from vegetables alone, three to six times the daily requirement of protein. The consumption by pygmy chimpanzees of fibrous foods every day, even when fruits are abundant, may be required to supply the necessary amount of protein.

Table 13. Amount of vitamin C and crude protein (fresh weight)
per 100 g of food (from Asato's unpublished analysis)

Food	Vernacular name	Plant part	Water content (%)	Pro-tein (%)	Vitamin C (%)
Fruits					
Landolphia owariensis	batofe	pulp	71.4	0.8	0.7
Mammea africana	bokoli	pulp	64.4	2.8	2.5
Cissus myriantha	bolengalenga	pulp	73.9	1.5	30.3
Croton haumanianus	bonyanga	pulp	74.5	2.8	3.8
Uapaca guineensis	bosenge	pulp	92.4	0.2	1.1
Dialium zenkeri	elimilimi	pulp	32.4	3.8	4.0
Dialium corbisieri	keke	pulp	34.0	2.6	3.3
Aframomum laurentii	ndake	pulp	96.1	0.2	4.0
Pancovia laurentii	tende	pulp	84.1	0.9	4.7
Beans					
Afzelia bipindensis	bala	bean	70.2	6.6	2.5
Anthonotha fragrans	boppembe	bean	38.0	6.1	0.4
Anthonotha macrophylla	lomuma	bean	45.2	5.1	1.3
Fibrous food					
Megaphrynium macrostachyum	beiya	shoot	88.9	4.2	8.3
Haumania liebrechtsiana	bokombe	shoot	90.4	3.2	1.8
Aframomum laurentii	bosomboko	pith	70.0	3.1	3.8
Leonardoxa romii	bokumbo	young leaf	75.0	3.4	7.3
Manniophytom fulvum	lokosa	young leaf	75.0	4.3	37.6
Palisota ambigua	liteletele	stem	79.2	4.0	2.7
Ancistrophyllum secundiflorum	bokau	pith	88.9	3.0	2.2
Cultivated fruit					
Ananas comosus	pineapple	pulp	84.9	0.4	14.1
Musa sapientum	banana	pulp	73.8	1.3	3.0
Carica papaya	papaya	pulp	86.4	0.7	79.2
Persea americana	avocado	pulp	83.9	1.2	7.1
Citrus sp.	A			0.0	70.3
Citrus sp.	B			0.0	28.1
Citrus sp.	C			0.0	29.5

The intake of animal protein seems generally to make only a trivial contribution to the nutrition of the pygmy chimpanzees of Wamba. The one exception is the larvae of the skipper butterfly (tohilihili). On average, one pygmy chimpanzee consumed approximately 200 caterpillars during each feeding bout, averaging 18 minutes. This feeding bout amounts to a fresh weight of approximately 100 g, which converts to 7.5 g of crude protein. Because the chimpanzees feed, on average, three times a day throughout the season, the calculated total quantity of crude protein for a day was 22.5 g, more than 60% of the protein required by a 35 kg adult. Although this value represents a meaningful contribution to nutrition, the food season of butterfly larvae ends after only a short time.

Grass-eating ruminants such as cows have a symbiotic relationship with protozoans living inside the alimentary canal. Hideo Hasegawa of Ryukyus University School of Medicine's Laboratory of Parasitology kindly allowed me to bring to him for analysis several hundred pygmy chimpanzee fecal samples from Wamba. He discovered in the feces a huge number of protozoans, along with eggs of various parasites. Because the protozoans were so big and their numbers so great, Hasegawa at first thought they were food fragments. I wonder if these protozoans have a symbiotic relationship with pygmy chimpanzees, but this is a problem for future examination.

Comparison with Common Chimpanzees

The number of common chimpanzee foods reported at Mahale, already 205 plant food items in 1974, presently amounts to 300 items. R. W. Wrangham (1977) reported 140 items at Gombe Park on the basis of two years of research, but the number exceeds 200 items when his observations are combined with those of other researchers. In the Ipassa Forest of West Africa, C. M. Hladik (1977) identified 174 items in one year, but if we include those things that are not yet identified, the number of items is 318. At Wamba, 147 food items of pygmy chimpanzees have been identified during research spanning ten years; this number is less than for common chimpanzees at any of the three above-mentioned regions.

Because differences in observation conditions or periods can bias data, it is dangerous to assert unequivocally that the common chimpanzee exploits a greater diversity of foods than the pygmy chimpanzee. Nevertheless, the tropical rain forest of Wamba is far richer and more complicated in floristic composition than the drier vegetation of Mahale and Gombe. Given the differences in the extent of food resources they can use, pygmy chimpanzees seem to use a relatively smaller portion than common chimpanzees.

A comparison of the number of food types eaten per month, as well as per day, also suggests that the common chimpanzee uses a greater diversity of foods than the pygmy chimpanzee. According to Wrangham (1977), 14.6 species per day and 60 species of food per month are consumed by common chimpanzees at Gombe Park. Hladik (1977) reports that an average of 20 kinds of food are used in one day by common chim-

THE LAST APE

panzees in the Ipassa Forest. Both are higher values than those obtained for pygmy chimpanzees at Wamba.

Food items, consisting of fruits, seeds, leaves, pith, bark, and various animal foods, are conspicuously similar in both species of chimpanzee. Among these items the proportion occupied by fruit (including seeds) in pygmy chimpanzees is 57% at Wamba, 63% at Yalosidi, 57% at Lomako, and in general, about 60%. In the common chimpanzee at Mahale, it is 34%, which is high when compared to the 6–10% of the gorilla, a fibrous-food eater, but low when compared to the pygmy chimpanzee.

Chimpanzees have been thought to be mainly frugivorous. Because fruit constitutes 80–90% of the diet at Wamba, estimated from the proportion of time spent feeding and from fecal analysis, pygmy chimpanzees are true to their reputation. In the common chimpanzee, however, the proportion of consumed foods differs somewhat from one site to another. At Gombe, the ratio of consumption according to feeding time was 47.1% fruits; about 32.0% leaves and bark; and 20.9% larvae and other material (Teleki, 1981). At Ipassa Forest, the proportion of fruit was still lower, a mere 14.4%, whereas the proportions of leaves and animals (larvae) were 48.8% and 36.8%, respectively (Hladik, 1977). At Mahale, the proportions of plant foods were calculated for two groups: fruits constituted 73–95% of the food eaten during one month by M-group and 7–47% of the food eaten by K-group (Nishida, 1974). Fruits comprised 90%, and other foods 10%, of the diet of common chimpanzees in the Budongo Forest (Reynolds and Reynolds, 1965).

Although there are exceptional cases such as the Mahale M-group and the Budongo Forest residents, in general, the importance of fruits in the diet of the common chimpanzee is lower than in the pygmy chimpanzee. Conversely, fibrous and animal foods are consumed in higher proportions by the common chimpanzee. In fact, a special characteristic of the common chimpanzee is that it ordinarily consumes animal food in notable quantities. A common chimpanzee at Gombe eats an average of 10 g of meat in one day and, in addition, may consume many invertebrates (Hladik, 1977). The chimpanzees at Mahale frequently feed on ants and also prey upon vertebrates; in 1981, 37 examples of meat eating were reported (Takahata, Hasegawa, and Nishida, 1984). It was also reported that the total time spent feeding on animal food amounted to 4% at Ipassa Forest. Compared to this, consumption of animal food at Wamba is extremely meager, amounting to no more than 1% of the total by fecal analysis and 1.5% by direct observations.

As previously stated, regional, group, and individual differences

] 114 [

occur in the pygmy chimpanzee diet, and are well documented for the common chimpanzee also. Probing into such dietetic differences at various levels may provide clues to the flexibility of the chimpanzees' food habits, in particular how they can change their food customs in response to environmental changes. Such probing may also help us to assess their food culture, in particular how they pass on their food customs from group to group, or change them from region to region. Presently, however, the data on regional, group, and individual differences are insufficient to compare the two species of chimpanzee.

One criterion for evaluating the plasticity of food habits is the ability to accept new kinds of food. The common chimpanzee eats many cultivated plants, among them sugarcane, papaya, sweet orange, and pineapple. Oil palm fruits are eaten by the common chimpanzees of West Africa and East Africa. Although the oil palm was brought to East Africa by way of Zaire, and thus must have an older history in Zaire, we have not yet observed pygmy chimpanzees using oil palms. The common chimpanzee may have the greater ability to accept new foods.

When comparing the food habits of the two chimpanzee species, we find that the common chimpanzee uses the habitat more intensely. It obtains a greater diversity of foods and is more progressive in acquiring new foods. That is, the common chimpanzee has a higher ability to exploit food. On the other hand, the pygmy chimpanzee consumes more fruit than the common chimpanzee.

Here, we must also consider the differences in food habits between the common chimpanzees of Mahale's M-group and K-group. Each group had a large, suitable, home range that overlapped the other in the Kasoge Forest. When M-group entered, however, K-group took shelter in the woodland belt (Nishida and Kawanaka, 1972). The proportion of fruits in the diet was high in M-group but low in K-group, perhaps the result of M-group's monopolization of preferred habitat. The proportion of fruits in the diet of common chimpanzees of the Budongo Forest is exceptionally high, but this region is known to be one of the highest density areas for the common chimpanzee.

The proportion of fruit in the diet becomes a point of speculation about the stability of diets. The pygmy chimpanzee, with its diet proportionately high in fruits, may have been in a stable food environment throughout its history. Conversely, the common chimpanzee, which is situated in a more diverse and unstable, severe environment, has developed a greater ability to exploit its food resources.

CHAPTER FIVE

Behavior of Individuals

Daytime Behavior

Although Wamba is situated directly on the equator, the exact times of sunrise and sunset vary by about 30 minutes during three months of the year. Coping with this time variation, pygmy chimpanzees leave their nests at between 5:20 and 6:20 A.M. and make their nests at night between 4:45 and 5:45 P.M. That is, the pygmy chimpanzees rise on average about 30 minutes after sunrise and retire about 40 minutes before sunset. The intervening 11.5 hours comprise the pygmy chimpanzee's daytime active period.

Because the pygmy chimpanzees live deep in the forest, they get up and go to bed in dim light. On mornings when there is rain or dense fog, they get up later, and after heavy rains they tend to retire early.

The daily activities of the pygmy chimpanzee are divided into six categories: feeding in the trees, rest, travel, foraging, nest-building, and group excitement. Because following individuals was inefficient, I based this classification on the movements of the whole party.

Arboreal feeding. During arboreal feeding, most members of a party are feeding in the trees. During the first 10 to 20 minutes after arriving at a feeding tree, chimpanzees feed excitedly and enthusiastically. Gradually, they calm down, become languid, and finally rest, but since behavior continually changes, clear boundaries between feeding and rest are difficult to observe.

Rest. During rests, most members of the party are inactive in the trees and on the ground or are indulging in characteristic rest-time friendly behaviors such as grooming or play. They also often make nests and nap. Seldom are all members resting at the same time; usually two or three individuals feed intermittently.

Travel. After a rest, one of the pygmy chimpanzees emits a "waa waa" in a loud voice, and the others cheer in chorus. This long distance call occurs intermittently, while the number of chorusing individuals increases. The party descends to the ground and the number of individuals displaying "branch-dragging" behavior increases. The choruses and displays increase in frequency, and pave the way for the forthcoming group-travel.

Following the last chorus, travel begins. Individuals who had been napping rush out from their nests. Individuals who had been grooming run several meters while shouting and then sit down. Then after a short rest, all at once, all of the members of the group begin to move, descending silently to the ground.

Pygmy chimpanzees usually travel silently on the ground until they arrive at the next food tree. Upon arrival, a loud chorus erupts, but while en route, not much vocalizing occurs. They usually cover a few hundred meters at a time, but on rare occasion they can travel as much as 2 or 3 km at a time.

Foraging. In the activity categorized as travel, definite direction and orderly movement are seen. Pygmy chimpanzees, however, frequently also travel very leisurely, without closely coordinated movements. One individual searches for food; another is absorbed in grooming and play; and another sits idly, or lies sprawled on the ground or in a tree close to the ground. The whole party, however, is slowly moving. For the moment, I am calling this type of group activity "foraging," which includes opportunistic feeding, moving, resting, and social activities. While foraging, the party is widely spread out, with the distance between the head and the tail of the party sometimes reaching 500 m.

Nest-building. In the daytime, party members build nests at different times according to personal preferences, but at night, all of the party members build nests more or less simultaneously. Upon arrival at their sleeping site, the chimpanzees, all at once, begin to climb the trees, and while giving the long distance call, begin to build their nests. No matter how large the party, all individuals finish making their nests, simultaneously, in 15 to 20 minutes.

Group excitement. There are various situations when the whole party enters into an excited state: when the chorus of a different party is heard; when a new party joins; or when a member of another unit group approaches. During group excitement, pygmy chimpanzees cry loudly,

A group at rest in the trees.

A pygmy chimpanzee dragging a branch.

Pygmy chimpanzees all moving in one direction.

scream in shrill voices, and run around in the trees and on the ground. They drag branches, and aggressive, reassurance, and sexual behaviors occur extensively. This state of excitement usually ends after ten minutes, but once in a while, it may last for half an hour or even an hour.

The daily activity of the pygmy chimpanzee consists of a cycle of resting, travel, feeding, and resting. The peak feeding hours are in the early morning, but when foods are not abundant, travel and food-gathering behaviors merge and become foraging behavior. After a second feeding peak in the late afternoon, the chimpanzees return to foraging, or to a combination of feeding and resting, and begin to travel toward the nest-building site.

The percentages of time spent in the various daily activities are, in order, 43% resting, 20% foraging, 20% feeding in the trees, 13% travel, and 13% other. Because some time spent on feeding is also included within the foraging category, I estimate that the time that pygmy chimpanzees spend feeding exceeds 20%, but does not exceed 30%, of the daily activity time.

Daily Travel Distance and Home Range

Our pursuit of pygmy chimpanzees begins upon arrival at their sleeping site before they have left their nests. Although they are easily followed while feeding in the trees, when they descend to the ground and begin to travel or forage, following them becomes difficult since they make few vocalizations on the ground. Nevertheless, as the evening nest-building time approaches, we can determine where they are from their many loud choruses.

Although following a party throughout the day is difficult, we note on a map the confirmed places from which the chimpanzees left their nests and the places where they made their sleeping beds at the end of the day. By rounding the distance covered in traveling and foraging to the nearest tenth of a kilometer (joining the points with a gently sloping curve), I determined that the median distance traveled during a day was 2.0 km (range, 0.4–6.0 km).

Most researchers would predict that the daily travel distance would be short when food was abundant and would increase during periods of food shortage. Although I observed this tendency, travel distances were not significantly different between food seasons. Differences in party size, however, were observed depending on the fruit season. When preferred foods were abundant, pygmy chimpanzees formed large parties.

During shortages between fruiting seasons, they tended to split into smaller parties. The ability to freely change party size, by the fission-fusion process, must have been adaptive since it enables effective use of food resources.

Until 1982, the reported home range of Wamba's E-group was 58 km². Of that total, however, 66% overlaps home ranges used by other unit groups. Some regions are even jointly held by three unit groups. The joint ownership of influential food resources by several groups is a distinctive feature of pygmy chimpanzees.

Because parts of primate home ranges are commonly used jointly by several groups of the same species, group home ranges are often divided into overlapping parts and a core area. The core area belongs exclusively to one group and includes several prominent food areas in which other groups are absolutely forbidden to trespass. The territory monopolized by Wamba's E-group, however, contains roads dotted with hamlets and consists of cultivated land and fallow secondary bush. As a food source, it is the least valuable part of E-group's habitat and it may be monopolized by E-group merely because the other groups neglect it.

Group ranges are not static. As the E_1 and E_2 subgroups have been drifting apart, so have the ranges of both groups gradually changed. At first, from 1980 to 1981, the E_1 subgroup extended its range to the west, resulting in an increase of the overlapping area with P-group. Moreover, starting in about 1983, the E_1 subgroup extended its range to the southeast, penetrating more than 1 km into the monopolized region of B-group. In the region of overlap is the swamp forest of *Uapaca*, a vast food zone that the E_1 subgroup now shares with B-group. Since 1981, the E_2 subgroup has extended its range to the north and northwest, and it travels southward into the range of the E_1 subgroup less and less frequently.

Habitat Utilization

As previously discussed, the vegetation of the Wamba region is generally classified into five types: dry primary forest, swamp forest, old and young secondary forest, and secondary bush (cultivated and fallow land). Table 14 is a summary of the rate at which pygmy chimpanzees use each vegetation type—that is, the number of days each vegetation type was used relative to the total number of days in the study, and expressed as a percentage.

Dry primary forest had the largest proportionate use, reaching 93.5% and indicating that it is the most important vegetation for the pygmy chimpanzee. The majority of the pygmy chimpanzee's principal

Table 14. Utilization of vegetation types (Kano and Mulavwa, 1984)

Vegetation type	Proportion of utilization in % (number of days used/number of days observed)				
	Dry primary forest	Swamp forest	Aged secondary forest	Young secondary forest and bush	Number of days
Landolphia owariensis	93.5	37.1	54.8	8.1	62
Transitional	100.0	50.0	50.0	0	2
Other Apocynaceae spp.	100.0	41.9	48.4	19.4	31
Transitional	100.0	12.5	43.8	31.2	16
Dialium sp.	97.1	42.9	20.0	31.4	35
Transitional	81.8	18.2	63.6	45.5	11
Pancovia laurentii	84.0	12.0	52.0	20.0	25
Transitional	77.8	11.1	55.6	11.1	9
Undetermined	100.0	25.0	62.5	37.5	8
Total	93.5	31.2	47.2	20.6	199

foods are produced in this vegetation type. Old secondary forest (47.2%) ranks second. Several important foods such as *batofe* (a fruit representative of the family Apocynaceae) and *ntende* (*Pancovia laurentii*, Sapindaceae) are produced both in primary forest and in old secondary forest. Herbs (Marantaceae), which constitute the bulk of fibrous foods, grow mainly in old secondary and dry primary forests.

The utilization rate of the swamp forest was 31%. The data collected in 1981 did not include the principal foods from the swamp forest, but in the latter half of December 1984, the principal species *bosenge-alosi* (*Uapaca hendelotii*) bore a large quantity of fruit in the swamp forest of the Luo River. The movement of the E_1 subgroup reflects dependency on that fruit for a full month. In addition to *bosenge-alosi*, several foods were seen in abundance in the swamp forest, such as the shoots of *mpeto* (*Sclerosperma mannii*), a species of Palmae, and African ginger (*Aframomum* sp.).

The utilization rate of young secondary forest and secondary bush was lower, averaging 21%. The utilization rate rose (31% to 46%) just before and after the *Dialium* season, when the fruit of the parasol tree (*Musanga smithii*) and the fruit and pith of African ginger were eaten. Chimpanzees are able to utilize these non-seasonal foods throughout the whole year and, in times of food shortage, their relative importance rises.

Quantitative data concerning the utilization of habitats across the vertical dimension have not been obtained. Nevertheless, because the overwhelming majority of principal fruits are produced in the forest crown, pygmy chimpanzees probably spend the highest proportion of

Table 15. Primates inhabiting the forests of Wamba

Common name	Vernacular name	Scientific name
Potto	*kachu*	*Perodicticus potto*
Bushbaby	*lisire*	*Galago demidovii*
Red-tailed monkey	*soli*	*Cercopithecus ascanius*
Mona monkey	*beka*	*Cercopithecus mona*
DeBrazza monkey	*punga*	*Cercopithecus neglectus*
Blue monkey (?)	*ikese*	*Cercopithecus mitis* (?)
(none given)	*ekele*	*Cercopithecus salongo*
Guenon	*tolu*	*Cercopithecus* sp.
Red colobus monkey	*yemba*	*Colobus badius*
Angolan colobus	*luka*	*Colobus angolensis*
Black mangabey	*gila*	*Cercocebus aterrimus*
Allen monkey	*elenga*	*Allenopithecus nigroviridis*
Pygmy chimpanzee	*elia*	*Pan paniscus*

their time in the upper tree layer. Some fruits (e.g., *ntende*) are also produced in medium-size trees. In the lower stratum, food production is poorest. Here, many trees do not produce fruits because they are saplings, and trees that do have edible fruits produce scantily. In search of food, the pygmy chimpanzee even extends its activities to the subterranean level, by digging earthworms, and into streams and swamp beds.

Many species of primates in addition to the pygmy chimpanzee have been observed at Wamba, and Table 15 lists them all. Although their habits have not been examined in detail, we believe that no other primate species uses such diverse food types and strata as the pygmy chimpanzee. For example, *Cercopithecus ascanius* occupies and is the master of the dry primary forest, leaving occasionally for secondary forest. The black and white colobus occupies the deepest part of the dry primary forest, and red colobus and DeBrazza monkeys are confined to the swamp forest. Compared with these sympatric primate species, the pygmy chimpanzee has the widest niche.

Searching for Food

Although a pygmy chimpanzee finds food at all levels of the forest, the principal source of nourishment is found at heights of 25 to 40 m above the ground. The feeding behavior of pygmy chimpanzees in trees occurs in three stages. First, a pygmy chimpanzee—let us say a male—climbs the tree to reach the food. Second, he gathers, transports, and ingests the food while remaining in the tree. Third, he descends the tree.

When the chimpanzee climbs a tall tree, which may have a diameter of more than 50 cm, he avoids a direct ascent, perhaps by using a vine dangling from a branch or twining around the tree trunk. When there is no suitable vine, he will use an adjacent narrower tree trunk to reach a bough of the taller food tree.

When climbing up, the chimpanzee's limbs move in the order right hand, left foot, left hand, right foot. The second through fifth fingers, which are very long, are used to hook or to clasp the substrate supporting the suspended body. The arms seem to be more propulsive than the legs, which are bent a little at the knee while climbing. If the substrate is thick, the feet extend to get a firm hold. If the substrate is narrow, they hold it lightly between the big toe and the other four toes, or, depending on the circumstances, they may clasp the substrate firmly. In this way, the position and movement of the foot against the substrate are similar to those

A pygmy chimpanzee climbing a slender vine, with an infant at its loins.

A branch bends under the weight of a young male pygmy chimpanzee as he brachiates, or arm-swings.

employed during horizontal movement in a tree. The ventral surface of the body does not contact the substrate.

Fruit is usually found at the surface of the forest crown, but the small branches at the tips cannot possibly support the weight of a pygmy chimpanzee. Consequently, when one walks along a big branch toward the tip, he or she stops just before a dangerous place, extends an arm, bends the branch, pulls it in, and removes the food. Also exhibited is a unique suspensory motion in which the body weight is dispersed by grasping at least two separate branches with the hands and feet. Then, before the branch breaks, the pygmy chimpanzee moves from branch to branch to alter the center of gravity, and approaches the food in the surface of the crown.

Pygmy chimpanzees eat only where they have a firm hold on the substrate. Therefore, they may have to carry the food a short distance. Pygmy chimpanzees usually carry food in their hands, but they also use their mouth or foot, though less frequently. Sometimes, they carry twigs or sugarcane in their inguinal region, a method first described by Jane Goodall as the "groin pocket" in the common chimpanzees at Gombe. At Wamba, certain individuals use it frequently, while others never do.

Pygmy chimpanzees usually sit when ingesting food. They sit most frequently with the soles of both feet on the substrate, a branch, and the buttocks or the base of the thigh on the same or another substrate. When the branch is slender, they contact the substrate only with their feet, and squat. Occasionally, they loosely dangle both legs while sitting on a branch, and they also bring food to their mouth while hanging by one hand. Although every imaginable variation occurs in sitting positions while eating, pygmy chimpanzees only occasionally eat while lying sprawled or standing quadrupedally, tripedally, or bipedally.

After feeding, pygmy chimpanzees rest and then descend to the ground to travel. When ascending to a feeding spot, they move up almost vertically, but when descending to the ground, they move diagonally. The average horizontal distance between the take-off point and the landing spot is 10 to 50 m. This horizontal displacement is related to their descent technique, branch-bending, which occurs whether they are moving quadrupedally or by brachiating. When a pygmy chimpanzee moves toward a branch tip, the branch bends because of his body weight, and he descends naturally. At the same time, the gap between him and the neighboring tree narrows. When the gap is still large, however, he may brachiate or leap across to the neighboring tree. Although pygmy chimpanzees will often descend using vines or trees, branch-bending is usually used at least

A male sitting and feeding on a *batofe* fruit.

It is not unusual to see a pygmy chimpanzee walk bipedally over a distance of 20 m while carrying sugarcane.

once per descent bout. Often individuals may even expressly climb to the top of a low tree in order to use branch-bending for a descent.

A pygmy chimpanzee seems to have an innate sense of how much a branch will bend under the force of its body weight. A pygmy chimpanzee is able to branch-bend by using a tree of appropriate thickness relative to the height of the dive. Once I saw a male, hurriedly following his party, dive toward the top of a lower tree from a height of 50 m. By repeating branch-bending three times in rapid succession, he landed on his feet on the ground in just seconds. It was an eye-opening aerial show.

Ordinarily, descent is slower. Especially before traversing a gap between trees, pygmy chimpanzees often take short rests of several seconds. Even when they descend a vertical tree trunk, they frequently come to a stop, still in descending posture, to take a short rest. This slow movement during the first stage of travel may provide time to gather relatives who had dispersed during feeding and resting, and to reach a silent agreement about the party's next direction and destination.

On the ground, pygmy chimpanzees knuckle-walk. Although Susman, Badrian, and Badrian (1980) say that they might use palmigrade walking (with the fingers spread and the palm down) in the mud of the swamp forest and in streams, I have always observed them knuckle-walking when on the ground, whether in a stream or on the forest floor. When they search for food on the ground, they may not need a high level of locomotor skill. They never use tools such as a piece of wood, and even when digging for worms, they only use their hands.

The branch-dragging behavior that occurs often on the ground is a special kind of locomotor behavior in which a pygmy chimpanzee moves in a tripedal run; the arm carrying the branch is slightly bent and held in the air. Less frequently, pygmy chimpanzees drag branches bipedally or quadrupedally with the branch in their mouth. They often move quadrupedally when carrying sugarcane in a hand or foot; they rarely carry it in the groin pocket or in their mouth.

Nest-Making

Nest-making behavior is common to all large-bodied apes. Under wild conditions, almost all individuals build nests in the evening and rest there at night. According to laboratory research, even zoo-bred chimpanzees who have never seen a nest will usually make something like one when given appropriate materials. They are never able, however, to make a complete one. Although chimpanzees have a natural disposition to nest-build, they also require experience and learning in infancy to complete the process (Bernstein, 1969).

Here I will examine from various angles chimpanzee nest-making behavior and the diversity of nests made by wild pygmy chimpanzees. This discussion will emphasize the flexibility of nest-building behavior, that is, the non-hereditary part.

Day nests and night nests. Basically, nests are made in the following way. A pygmy chimpanzee stands on the fork of a branch with all four feet and folds the surrounding branches inward, stamping down on them. When the external appearance of the nest becomes almost spherical, the chimpanzee sits in the hollow in the center or lies sprawled in it. After

The alpha male of E₁ subgroup, Ten, resting on his crude day nest.

A pygmy chimpanzee getting up from his night nest in the morning mist. A night nest is larger and more sturdy than a day nest.

A male resting on a day nest.

Pygmy chimpanzee in a day nest.

An old nest of a pygmy chimpanzee.

completion of the basic nest, the occupant often bends down small twigs projecting into the interior of the nest, breaks off small branches with leaves attached, and adds various elements for cushioning. The average size of the eliptical nests was 90 cm long by 70 cm wide.

Pygmy chimpanzees make nests at night for sleeping, but they also frequently make nests during the day for naps or a rest. Night nests and day nests seem to differ in several ways. First, the nest-building sites differ. Day nests are often made opportunistically in trees where the pygmy chimpanzees are feeding, and consequently, day nests are built in secondary forest as often as in primary forest. In the afternoon, pygmy chimpanzees frequently enter secondary forest in search of *beiya* and other ground foods, but as evening approaches, they invariably return to primary forest to build night nests. Pygmy chimpanzees will often travel as far as 500 m just to build a night nest; I have wondered why they are so attached to the dry primary forest for sleeping at night. The attachment may relate to the distribution of preferred nest trees. Although 108 species of trees have been recorded as being used for night nests, one tree species among them, the *bokumbo* (*Leonardoxa romii*), occupied more than 39% of the total 3,353 nest trees recorded. The ten top-ranking species occupied 72% of all trees (Table 16). Most of the tree species that were preferred for nesting occur mainly in the dry primary forest. Within the primary forests of the E-group range, several areas are used very frequently as sleeping sites. These areas are densely populated with nesting

Table 16. Nest trees

Tree species used for nests	Number of nests recorded	Percent
Night nests		
Leonardoxa romii	1,311	39
Duvigneaudia inopinata	207	6
Scorodophloeus zenkeri	194	6
Uapaca guineensis	140	4
Garcinia epunctata	134	4
Anonidium mannii	114	3
Cavacoa quintansii	100	3
Pancovia laurentii	85	3
Cleistanthus mildbraedii	72	2
Polyalthia suaveolens	65	2
Other (98 spp.)	931	28
Total	3,353	100
Day nests		
Scorodophloeus zenkeri	89	14
Dialium spp.	77	12
Uapaca guineensis	69	11
Leonardoxa romii	67	11
Anonidium romii mannii	34	5
Anthonotha macrophylla	26	4
Gilbertiodendron dewevrei	25	4
Pancovia laurentii	23	4
Duvigneaudia inopinata	19	3
Ochthocosmus africanus	19	3
Other (56 spp.)	189	30
Total	637	101

trees of the preferred species, and the part of the dry primary forest that is less frequently used as a sleeping site has a low density of the preferred nest trees.

By contrast, not much selectivity is exercised in the choice of day nest trees. In fact, choice appears to be random. Sixty-six species are recorded as day nest trees ($N = 637$ nests), but of that number, 13 species have not been reported even once as night nest trees.

Several other differences are seen between night and day nests. The average time for nest-making at night was 222 seconds (range, 96–412 seconds; number of nests measured, $N = 50$), compared to 42 seconds (range, 3–241 seconds; $N = 89$) for nest-making during the day. Time was measured from the beginning of nest-making until the shape was complete; that is, from bending down the first branch until the chimpanzee sits down or sprawls in the nest.

It is difficult to distinguish night and day nests by their shape and size, but on the basis of the length of time it takes to frame a nest, the night nests are made much more carefully.

Differences are seen in nest form, and I classified nests into three basic forms, A, B, and C, according to the location of the nest in the tree (Fig. 10). Form A nests were located at the top of an upright, often low, tree. Form B nests were located at the tips of side branches of medium and tall trees. Form C nests were located on the basal or medial part of the branches of medium to tall trees, closer to the trunk than the B form. Form A nests pitch easily from side to side, and form B nests pitch easily up and down and from side to side. In contrast to the instability of forms A and B, form C nests are generally firm at their base.

A large number of nests are made by using two or more trees. In those cases, I classified nests as combinations of the three basic forms according to the position the nest had relative to each tree, calling them for example, form AA or form ABC (Fig. 10). The maximum number of trees that contributed to one nest was six ($N = 2$), form AAAAAA and form AAABBB. In addition to the three basic forms was a nest type I call form D. This nest form was made on horizontal vines and, without exception, appeared in combination with other forms. Only combination nests that include form C have a stable base.

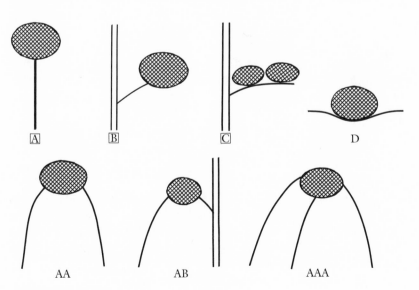

A   B   C   D

AA   AB   AAA

Figure 10. Nest forms (positions of nests in the trees).

Form B and the combination forms that included form B were the most numerous, whether in day or night nests. On the other hand, form C and its combination forms were recorded in only 15% of all night nests, while they occupied a considerably higher percentage, 38%, of all day nests (Fig. 11). Generally, combination forms are found at a higher rate in night nests than in day nests.

Many night nests were constructed in medium to low trees, but height selection did not seem to occur in the framing of day nests. Consequently, 59% of night nests were made above the ground and below 15 m, but 68% of day nests were constructed at a height above 15 m (Fig. 12).

Chimpanzees occupy day nests an average of only 17 minutes (range, 0–194 minutes; $N = 293$). The nests of form C, made simply with a firm base, are hard underneath and may not be very comfortable for sleeping. In contrast, a night nest must last for more than 12 hours and requires, simultaneously, stability and comfort; therefore, selectivity is seen in the place, kind of tree, and branch used, and night nests are built more carefully. Night nests also tend to be built in lower places than day nests (Fig. 12). There may be more trees or branches suitable for making night nests in the medium to low layers than in the tall layer, and/or the chimpanzees who sleep in lower places may be more comfortable at night than those at higher places because the vegetation at lower heights is thicker and may shelter them from night rain more effectively.

When resting during the day, a chimpanzee often approaches and gets into another's nest. While sitting on the rim, the visitor may begin to groom the resting one. A visitor may also add branches to the nest of another individual, and after the nest is enlarged, sometimes the individual gets in. Often, sexual behavior takes place in a nest, and on rare occasions, two individuals cooperate in making a nest. However, it is unusual for night nests to be used as a place for social life.

Often a mother will make a day nest and let her small infant play in it. While lying sprawled looking up, she will tickle the infant and hold its hands and feet; hanging high in space, the infant looks very happy and fortunate. If the mother happens to fall asleep, the baby may move around inside the nest, clamber around the edge, and peep out. Because the day nest serves as the infant's shelter as well as her own, a mother makes the nest very carefully compared to a male (mean nest-construction time is 85.5 seconds for females vs. 37 seconds for males). As stated above, the nest is made differently depending on the purpose and circumstances.

Regional differences in nests. The pygmy chimpanzees of Yalosidi and Lake Tumba are known to build nests on the ground, but at Yalosidi es-

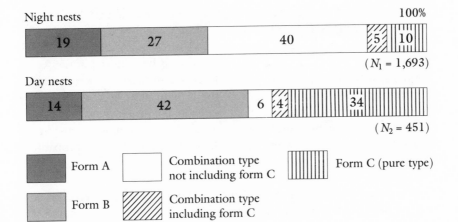

Night nests

| 19 | 27 | 40 | 5 | 10 |

(N₁ = 1,693)

Day nests

| 14 | 42 | 6 | 4 | 34 |

(N₂ = 451)

Form A   Combination type not including form C   Form C (pure type)

Form B   Combination type including form C

Figure 11. Proportions of nest forms.

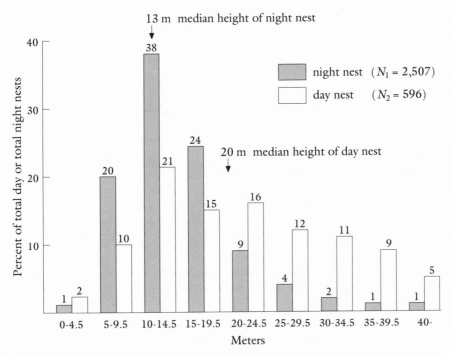

13 m median height of night nest

night nest (N₁ = 2,507)

day nest (N₂ = 596)

20 m median height of day nest

Figure 12. Relative heights of day nests and night nests from the ground.

pecially, ground nests accounted for 7% (71 of 1001 total nests) of those recorded during 1973 (Kano, 1983). At Yalosidi, there were two forms of ground nests. In the first, which predominates, a ground cushion is formed when shrubbery on the circumference is bent inside; in the second, branches that had been broken and carried from somewhere else are arranged on the ground. Almost all the ground nests at Yalosidi were found in dry primary forest. On one occasion, ten new ground nests and many tree nests were found in proximity. All were fresh and seemed to have been made at the same time. These ground nests were probably built for night resting, because day nests are rarely made in large numbers at one site.

In contrast, although several thousand nests have been recorded at Wamba, only a few ground nests have been seen. Similarly, ground nests have not been reported in the Lomako Forest. From the west bank of Lake Tumba, Arthur Horn (1980) reported four cases of ground nests of the second form; whether those ground nests were made for day or for night resting is not known.

At Wamba and Yalosidi, nests resembled each other in terms of the height of the nest above the ground and the tree species used. In both areas, many night nests were made in relatively low places (above ground and <15 m) and the preferred nest tree was *bokumbo*. According to my survey, *bokumbo* was preferred as a nesting tree for the pygmy chimpanzees not just at Wamba and Yalosidi but throughout the extensive areas covering the upper Tshuapa basin.

At Lomako, results were somewhat different. Alison and Noel Badrian (1977) reported that the pygmy chimpanzees of Lomako had a strong tendency to build night nests in the trees where they fed. At Wamba, however, pygmy chimpanzees rarely made nests in feeding trees except during the afternoon rest period. At Lomako, night nests were made in relatively high places, with more than 50% of the nests being made more than 20 m above ground level. In this respect, Lomako clearly differs from Wamba and Yalosidi. My data show that, where human hunting pressure is high, pygmy chimpanzees build their nests in high places. According to the Badrians, however, human hunting pressure against pygmy chimpanzees at Lomako is low.

Why so many ground nests are at Yalosidi is puzzling. I am unable to show a relationship between ground nests and numbers of predators. At Lake Tumba levee, where ground nests are seen, hunting pressure against chimpanzees is very high. At Yalosidi, there are also many hunters, but pygmy chimpanzees are not their target. The density of chimpanzee predators such as leopards may be higher at Yalosidi than at Wamba.

Pygmy chimpanzees exhibit a variety of nest-making behaviors, probably in response to various ecological conditions. Regional differences in nest building also exist, and future research should reveal the causes of these variations.

Comparison with Common Chimpanzees

Pygmy chimpanzees use almost every vegetation structure in their home range. They also use every stratum of the forest. Because the middle and low strata, little used as food sources, are most frequently used as nest building sites, well-balanced use of each forest stratum is realized. Moreover, the pygmy chimpanzees use the underground, streams, and swamps in the forest.

Common chimpanzees also make wide use of available resources. For example, the chimpanzees of West Tanzania utilize patches of tropical rain forest (small-scale rain forest isolated like a small island), riverine forest, montane forest, regenerating forests, woodlands, open acacia savanna, and grassland—every possible habitat type within the home range—as food sources and sleeping places (Kano, 1972). In fact, the ability to use the environment comprehensively may be the distinctive feature of the genus *Pan*. Both species of chimpanzee, relative to other nonhuman primates, may have retained "generalized" features and, consequently, are not limited to a particular habitat.

The two species of chimpanzee differ in habitat use in that the pygmy chimpanzee occupies a somewhat humid habitat and the common chimpanzee occupies a dry one. Pygmy chimpanzees, unlike common chimpanzees, are not afraid of the water. Swampland bogs and small streams serve as their daily feeding place. By contrast, common chimpanzees typically have a morbid fear of the water. Even when crossing shallow small streams, they make a great effort, by using every rock and tree, to keep their body out of the water. Actually, they do not seem to know how to deal with water, and one can drown even in a small pool of water inside a zoo cage. The common chimpanzee, however, appears to be much better adapted to dry regions than the pygmy chimpanzee, as shown by the common chimpanzee's advancement on to the savanna and grassland.

The two species differ also in population density and home range size. The group density of 1.7 individuals/km^2 (and the home range size of 58 km^2) observed for pygmy chimpanzees at Wamba is the highest density zone of the pygmy chimpanzee. Nevertheless, this is lower in density (and larger in home range size) than that of the highest density zone of the common chimpanzee (Table 17). At the very dry marginal habitat,

Table 17. Size and density of chimpanzee home ranges

Chimpanzee group	Group size	Population density	Reference
Pygmy chimpanzees			
Wamba	58	1.7	Kano and Mulavwa, 1984
Common chimpanzees			
Gombe	26	2.6–3.9	Teleki, 1973
Mahale (K-group)	10.4	5.7	Nishida, 1979a
Mahale (M-group)	13.4	5.7	Nishida, 1979a
Budongo	20[a]	3.9[b]–6[c]	
Bossou	5–6	3.5–4.2	Sugiyama and Koman, 1979
Ugalla	700–750[d]	0.08[e]	

[a]Suzuki, 1977 [b]Reynolds and Reynolds, 1965 [c]Sugiyama, 1977 [d]Itani, 1979 [e]Kano, 1972

however, the density of common chimpanzees is extremely low, 0.07–0.08 individuals/km² (Kano, 1972). According to J. Itani's (1979) estimate, in these regions, the home range of one unit group amounts to 700 to 750 km², which may be close to the range of a Bushman band.

We do not know the lowest density at which pygmy chimpanzees can survive and maintain normal reproductivity in natural conditions, because their extremely low density in many present localities is presumed to be caused primarily by human pressure. On the basis of the uniformity of vegetation conditions, however, they appear to have less flexibility in density and home range area than common chimpanzees. That is, common chimpanzees seem to possess a flexible nature enabling them to endure more varied environmental fluctuations.

The common chimpanzees of Gombe Park spend more than 40% of their daily activity budgets feeding (Teleki, 1981). Moreover, this high rate is maintained throughout the dry-wet seasonal cycle. Relative to common chimpanzees, pygmy chimpanzees spend less time feeding, and consequently, relatively more time resting, which includes various other social behaviors. Thus, common chimpanzees may expend more effort for individual maintenance than pygmy chimpanzees.

Pygmy chimpanzees may surpass common chimpanzees in mobility in the trees. The width of the shoulder blades is narrower and resembles that of a gibbon. Adriaan Kortlandt (1967) stated that "the pygmy chimpanzee is a gibbonized chimpanzee." Although field observations indicate that pygmy chimpanzees may, in fact, be more agile in the trees than common chimpanzees, both species move in basically the same way. In the trees, they generally walk quadrupedally when moving horizontally. Unlike gibbons, pygmy chimpanzees do not perform long-distance brachiation and, furthermore, they rarely exhibit the gibbon's acrobatic skill at ricochetal brachiation, in which the animal springs forward with each arm-swing. Rather, pygmy chimpanzees often move by bending a branch using their large body weight.

Another difference between pygmy and common chimpanzees is that common chimpanzees use tools when feeding. For example, common chimpanzees pick shrub and grass stems of a suitable length and use them for collecting termites or ants (Goodall, 1965, 1970; Nishida and Hiraiwa, 1982). When they find an acceptable tool, they may even transport it 800 m to the next termite mound. There are other examples. West African chimpanzees, which eat not only the flesh but the endosperm of the oil palm nut, must first crack the walnut-like hard covering. They place the nut on a stone and hit it with a special rock, cracking the nut shell; then they pick out and eat the endosperm (Sugiyama, 1979). In zoos and in the field, common chimpanzees are reported to use pieces of wood to reach food when a branch bearing fruit is beyond their grasp. Also, when capturing small animals, common chimpanzees seem to know to work cooperatively to corner and run them down.

Pygmy chimpanzees, in contrast, do not use tools when feeding. In the rain, at Wamba, a pygmy chimpanzee may pick off a small branch with the leaves attached and put the branch over its head and back (Kano, 1982b). The same kind of behavior has been reported in another large-bodied ape, the orangutan (Rijksen, 1978). Because the small branch with attached leaves was used as a raincoat, this might be regarded as one kind of tool-using behavior. There are, however, no reports of pygmy chimpanzees using tools to procure or process food. Badrian, Badrian, and Susman (1981) mentioned that a stick thrust into a termite mound at Lomako might have been the act of a pygmy chimpanzee, but they could not confirm it by direct observation.

In summary, common chimpanzees exploit a wider ecological range to obtain food than pygmy chimpanzees and have developed a higher degree of technical skill. Conversely, the food acquisition skills of pygmy chimpanzees are primarily for obtaining fruits high in trees. The short time invested in feeding activity suggests that they do not occupy as severe a food environment as common chimpanzees.

CHAPTER SIX

Sexual Behavior

The Role and Diversity of Sex

In primates, males and females usually live in the same group. The famous Sir Solly Zuckerman claimed that "sex" was what allowed males and females to coexist in these primate groups (1932). His claim was based on the year-round copulation of baboons at a zoo. Because many tropical primates do not have fixed breeding and birth seasons (and because all primates are natives of the tropics) Zuckerman's assertion was accepted for a long time.

This assertion, that sex was the factor holding primate groups together, was eventually overthrown by research on Japanese monkeys (Lancaster and Lee, 1965). From November to February, adults mated, and from June to August, infants were born. In Japanese monkeys, which experience four severe seasonal shifts, the breeding season is fixed, and infants are born in the period when child-rearing is easiest. Although copulation does not occur at all during half of the year, Japanese monkeys have formed very stable multi-male groups. Thus, factors in addition to "sex" must be involved in the maintenance of this stable social structure. The assertion, however, that social structure is not at all affected by the influence of sex is too brash. Considering that reproduction is strived for within groups that include males, females, and offspring, some form of sex is undoubtedly involved in the maintenance of society.

The two species of chimpanzee do not have special mating and birth seasons. Though the two are basically identical in their sexual physiology, they nevertheless display great differences in sexual behavior. Below I consider whether those behavioral differences are reflected in differences in the social structures of the species.

The term sexual behavior includes more than reproductive activities, especially in higher animals in which the steps leading to copulatory behavior are often complicated. A general example is the courtship display of a male toward a female; the pattern of courtship display in each species of animal is unique. Some behaviors originally unrelated to sex may take on sexual meanings. For example, C. R. Carpenter (1942) discovered that grooming plays an important role in the mating activities of the rhesus monkey, and as such, he called it a secondary sexual behavior.

Conversely, some behaviors thought to have developed from reproductive behavior have lost their original meaning, and now function only as social behaviors. Male mounting, seen extensively among primates, is a good example. Although similar in form to copulatory behavior, male mounting, as seen in baboons and macaques, is a nonreproductive social behavior that is related to dominance and other social activities. In the pygmy chimpanzee, behaviors, such as mounting, genito-genital (GG) rubbing, and rump-rump (RR) contact, that involve contact between external genitalia are categorized as behaviors that have their origin in copulatory behavior. They do not now have reproductive functions, but because they make such a vivid sexual image, we could probably call them sexual behaviors.

In this way, sexual behavior is a very broad category, and the boundary with social behavior is not distinct. Among the behaviors that appear to be sexual, however, we shall ignore the particular behavioral form and pay attention to the functional aspects only. That is, I shall accept as sexual behaviors only mating behaviors that ought to have reproductive functions—sequences of behavior leading to copulation (for example, courtship displays) and behaviors that conclude mating (postcopulatory behaviors). Consequently, mounting (which has the appearance of copulation but does not have a reproductive function), GG rubbing, RR contact, and other "mating-like behaviors" will not be regarded as sexual behavior. These will be investigated in the next chapter.

Similarly, because "immature" copulation between an immature individual and a mature individual or another immature individual does not have a reproductive function, this behavior is not treated here. In this chapter then, I address only "mature" matings between postadolescent males and females thought to be sexually mature, and "copulation" will always indicate "mature copulation."

Copulation

During the period from October 1975 to February 1979, 219 examples of copulation under natural conditions were reported (700 observation hours) and 515 examples under conditions of provisioning (330 observation hours).

The courtship displays of pygmy chimpanzees, like many animals, are performed by a male and directed toward a female. There are various patterns of courtship displays. In the first form, a male attracts the attention of a female who is far enough away that she cannot be touched, faces her, and spreads his thighs. While exposing his erect penis, he sits or moves his squatting body up and down, forward and backward, or side to side. Some males stick out their chest, and some stoop forward while they are displaying. Their hands are extended and moved in any direction according to individual preference, upward, sideways, and, most often, forward, as if to beckon the female.

If the female does not respond to this first form of courtship display, the male often changes to a second form of display. He approaches the female, sits or stands bipedally, and extends his hand, lightly touching and then releasing her head, shoulder, back, or knee. His body slowly moves forward and back and side to side. One male moved his body violently up and down in the first display form, but when the female approached, he slowed his movements and touched her body cautiously and gently as if touching a fragile article.

A female responds positively to a courtship display by standing when the male is near or by standing up and approaching the male. After this, she exhibits "presenting" behavior, in which she presents her genitalia to the male; he responds by mounting and copulating with her. If a female does not exhibit presenting behavior, the male retreats to a place separated from the female by several meters and resumes the courtship display. Repetition of the courtship display, followed by the female's approach, the male's retreat, and the next courtship display, recurs several times. We often saw a male draw a female away from other individuals, go up into a tree, and copulate with her there.

Females also solicit copulation, approaching and presenting to males even though they have shown no sign of courtship. Of copulations observed in their entirety during 1984, 71 were examples in which the male approached or first made a courtship display and 25 were examples in which the female approached and solicited copulation.

Although the majority of copulations were scored as having been initiated by males, we are not certain that the initiative was always actually taken by the male. Often an estrous female would enter the feeding site, approach an unsuspecting male, and choose a feeding place not too close, but not too far, from the male. This would be, for humans, like casting an amorous glance. At these times, the male chimpanzee might be fascinated by her swollen genitalia. Although the female would appear not to notice the male at first, when he showed a little courtship display movement, she would suddenly approach and present to him for copulation. In this situation, we cannot say that the male initiated copulation.

"Peering behavior" may be another kind of precopulatory behavior. Because it occurs in various social contexts that do not involve partner responses, the goal of the actor is often unclear. Peering rarely occurs in sexual contexts, but may in two cases. A young male whose repeated and persistent displays are ignored by a female may begin to peer at her. Also, a young female may peer at a male while she is begging for food from him, and he may respond by copulating with her.

In pygmy chimpanzees, every copulation is invariably preceded by female presentation, which takes two forms. In one form, the female stands quadrupedally, facing away from the male and thrusting her genitalia in front of the male; this is followed by dorso-ventral copulation. In the other form, the female lies on her back in front of the male, spreads her thighs, raises her buttocks, adjusts herself, and shows her sexual organs; the male approaches and copulates ventro-ventrally. Both body positions are regular copulatory positions, but the dorsal one is more frequent.¹ Ventral copulations under natural conditions were 29.1% of the total copulations reported ($N = 179$) and 7.6% of those reported at the feeding site ($N = 487$). These results on copulatory positions in the field differ from observations of captive pygmy chimpanzees. At the Yerkes Research Center, the ventro-ventral position was assumed in 53% of the cases (Savage-Rumbaugh and Wilkerson, 1978), and at the San Diego Zoo, only the ventro-ventral copulatory position was observed (Patterson, 1979). \

The choice of copulatory position differs according to age. Ventral copulation is more frequent in both male and female adolescents than in full adults. The external genitalia of a young female pygmy chimpanzee swell and protrude backwards slightly. With age, however, they become large, develop a unique shape, and hang between the thighs. Tom Patterson (1979) suggested that the pygmy chimpanzee takes a ventro-ventral copulatory position because of this peculiar placement of the sexual organs, which point and extend toward the thighs in older females; inserting the penis is easier in a ventral than in a dorsal position. Observations in the wild, however, do not confirm this explanation. In the wild, the dorso-ventral position is more frequent. Furthermore, young females, whose sexual organs do not face ventrally, choose ventro-ventral copulation more frequently than do older females.

Although the percentage of ventro-ventral copulations was low overall at Wamba, half of the observed copulations of a female with an immature male were ventro-ventral. The pygmy chimpanzee colony at Yerkes consisted of a young female, one-half year after capture, and a juvenile male. Because the copulations occurring between them were "immature" copulations, no real discrepancy exists between the Yerkes colony and the Wamba group. Both individuals observed at the San Diego Zoo were adults, but they were infants at the time of their capture. (The observations at the San Diego Zoo suggest that socially deprived pygmy chimpanzees may not develop normal sexual behavior, but they do not prove that pygmy chimpanzees normally take the ventro-ventral copulatory position. /

Some say that males prefer the ventro-ventral position to the dorso-ventral position. According to E. Tratz and H. Heck (1954), the male pygmy chimpanzee at the Frankfurt Zoo lightly touches the female's body with his hand and induces her to present ventrally in front of him, after which they copulate. This alleged preference also contradicts field observations. In the wild, when a female presents ventro-ventrally, frequently the male will not want to mount, whereas in dorsal presentation, this unwillingness is rarely observed. In contrast, several cases of the following type have been reported: the female presents dorsally, but immediately before the male mounts her, she turns her body over and forces a ventro-ventral copulation by embracing him ventro-ventrally. Similar changes have also been seen during copulation. While in a dorso-ventral position and with the male in the midst of thrusting vigorously, the female will suddenly disengage and lie on her back; with both legs, she grasps the male's hips and changes to a ventro-ventral copulatory position. Changing of body position during copulation has been reported a total of 16 times, but all of the cases were changes from a dorsal position to a ventral one, not the reverse. In the wild, the ventral position is clearly preferred more by females than by males.

There are few variations in the posture of ventro-ventral copulation. The female passes both of her arms under the male's arms or she turns on her shoulder and clings to the male; with each of her legs, she grasps the male's hips. Occasionally, instead of embracing the male, she will grip an overhead branch with both hands or with one hand. The male's posture is quadrupedal. Often, the female lifts the middle of her back from the ground or branch, but there are also times when her back touches the support structure. In all ventral copulations, at least part of the female's weight seems to be supported by the male.

The dorsal copulatory posture is richer in variations. A female will

An adolescent female presents her genitalia ventro-ventrally to an adolescent male.

The male mounts the female.

The male achieves intromission and begins pelvic thrusting.

often take a quadrupedal stance with slightly bent limbs; she may also, however, lie completely prostrate on her belly, take a posture with only her buttocks jutting upward, or stand with all four limbs taut. There is also "sitting copulation," in which the female sits with her back toward the male. Occasionally, while facing sideways or lying on her back, she may also even engage in "lateral copulation," in which both her legs are twined around the side of a male's chest. This latero-ventral copulation is intermediate between the ventro-ventral and dorso-ventral positions, but for convenience, I will treat it as a dorsal position. Normally, a male assumes postures varying from a bipedal standing slouch to a sitting position. He often puts his hand or his elbow on the back of a female, but sometimes he lets go with both hands.

During dorso-ventral copulation, females frequently extend a hand or foot between the male's legs to touch the bottom of the scrotum. The variations on this theme are great, from pressing against the scrotum to intensify the pressure of the thrusting movement to not really touching the scrotum at all in a merely perfunctory manner. A female in the same posture will occasionally push the lower part of her genitalia upward as if to increase the friction of a full thrust of the male's penis.

An average copulatory bout (from the male's mount until his dismount) lasted 15.3 seconds at the feeding site. During that time, the male maintains a rapid, rhythmical thrusting motion. From observations of 8-mm film, the average number of thrusts per second and per copulation were 2.7 times and 43.8 times, respectively.

Whether or not ejaculation occurs during copulation is difficult to determine, and the female's vaginal plug, formed by the white ejaculate, is rarely seen. For two reasons, however, I think that ejaculation during copulation is not common. The first reason is that after copulation the male's penis is frequently still erect. The second is that the number of copulations per day per male is great. It is not unusual for the same male to copulate several times in one day. At the feeding site alone, the same individual was seen copulating 11 times, the maximum number of copulations per day recorded for one male. At the feeding station, the total number of copulations by males who copulated twice or more in one day was 252, approximately half the number of all observed copulations. Because some copulations must occur outside the view of the observer, males probably engage in many copulations every day.

Many copulations happen silently, but also commonly, the female screams when a couple approaches the end. This scream may relate to ejaculation and female orgasm, because often when the same couple copulates repeatedly in a short interval, the scream is more likely to be heard during the last copulation. All of the copulatory screams of female partners, however, are not related to ejaculation or female orgasm. For example, I observed a case in which a female gave a scream while copulating with a male who had no external sexual organs except a stunted penis.

After copulation, one or both partners remain sitting in the same place or separate from each other by several meters. Very rarely, after a male dismounts a young female, she gives a scream and runs several meters away. Usually, however, both partners are calm after copulation.

Reactions of Others

Pygmy chimpanzees are promiscuous. All kinds of pairings among group members are possible, except between a mother and her mature son. Consequently, members of the same sex within a group can all be sexual competitors. At the feeding site, however, interference by other mature individuals before, during, or after copulation was seen in only 33 out of 515 copulations. That is, interference during copulation occurred at the inconsequentially low rate of 7%. Adult pygmy chimpanzees apparently are generally tolerant of each other's mating activities. We cannot assert this unconditionally, however, because a male may intentionally solicit and copulate with a female who is separated from other members precisely to avoid interference from other individuals.

Twenty-seven of 33 examples of interference were by adult males; the other 6 cases were by adult females. There were 11 cases of approach, 2 cases of intimidation, 8 cases of threat and pursuit, 10 cases of slapping with the open hand, and 2 cases of squeezing in between the partners. When the interferer was a male, he was without fail dominant over the copulating male, who inevitably fled or avoided the interferer upon his approach. When the copulating female was young, the new couple fled together, but in general, females were calm in the face of interference. In fact, some even behaved aggressively toward the obstructor.

The following three examples illustrate some of the variety of responses to interference during copulation.

EXAMPLE I. 8:30 A.M., February 4, 1979. The second-ranking male Yasu approached the fourth-ranking male Hachi while Hachi was copulating with a middle-aged female Chiyo, and Yasu attacked Hachi. Hachi was confused, and when he dismounted from Chiyo's back, Chiyo savagely leapt upon and clung to Yasu. Yasu seemed startled by this unexpected counterattack and fled several meters. He then went back to Chiyo, exhibited behavior something like a sexual display, and, in a soothing way, mounted her.

The next two are examples in which a female behaved aggressively toward her sexual partner who either dismounted without finishing copulation or did not respond appropriately to the female's solicitation.

EXAMPLE 2. 6:50 A.M., January 13, 1978. When a young male Jess began to copulate with the young female Mayu, Yasu, who was several meters away, threatened Jess. Jess immediately got down from Mayu's back. Mayu savagely attacked Jess and drove off Jess and Tare, a middle-aged female who was a short distance from Jess. Jess screamed as he fled.

EXAMPLE 3. 1:50 P.M., February 11, 1978. When the third-ranking male Ude came out to the edge of the feeding site, the old female Sen approached and presented to Ude. Ude hesitated a little, and after pressing his buttocks lightly against Sen's genitalia (RR contact), he continued on to the feeding site. Sen screamed furiously and attacked Ude. Ude, who was annoyed, fled from place to place, but Sen pursued him obstinately and continued the attack while shrieking. Meanwhile, the other females joined in, creating an uproar at the feeding site. Ude at last climbed a tree and escaped disaster.

The above three examples show that females are active or assertive in sexual interactions with males and, at the same time, that the dominant-subordinate relations between males and females are not clear and invariable. High-ranking males usually receive appropriate respect from lower-ranking males, and certainly from females. Females, however, can get irritable, and if they do, even high-ranking males cannot touch them. At these times, pygmy chimpanzee society resembles a society of hen-pecked husbands.

Copulation and pseudocopulatory behavior (GG rubbing, RR contact, and mounting) appear to stimulate the sexual drive of other individuals. Interference can also stimulate sexual activity, but more commonly a female or male will look patiently for another partner and enter into sexual interactions when a partner becomes available.

Juveniles show the strongest interest in the performance of copulation. They run up to watch, cling to the belly or back of either partner, and scream. Often, they are the children of the female partner, because a child is usually situated close to its mother. But, if copulation by a female other than the mother occurs nearby, the juveniles will hasten to that

It is not uncommon for an immature to get between mating partners, and the behavior is tolerated by the adults.

A juvenile comes between a male and a female during copulation.

The juvenile screams while clinging to the copulating male.

Multi-individual copulation: a male, A, mounts and thrusts against one of two females who are genito-genital rubbing. Another male, B, approaches the group from the left.

Male B embraces male A dorso-ventrally while male B thrusts against the GG-rubbing females.

place in the same way. Their scream is not from fear or distress, but instead seems to come from sexual excitement; often after copulation, the child will enter into sexual intercourse (immature copulation) with one of the partners.

This kind of open participatory response to the stimulus of sexual relations of others is occasionally seen in sexually mature individuals. The term "multi-individual copulation" applies to such behaviors that maintain the sexual interactions (copulation and pseudocopulation) of others. In response to a case of copulation and mounting, a second male might mount the male who was mounting, and thrust. In response to GG rubbing, the male would mount the female who was on top, insert his penis, and thrust. These multi-individual copulations may be an extension of the sexual behavior characteristic of juveniles and infants. Such behaviors are infrequent and seen primarily in males during adolescence, and they disappear when adulthood is attained.

The Development of Mating

The majority of copulations of pygmy chimpanzees occur under a limited set of conditions, usually in the early morning. Under natural conditions, 69% of 219 copulations were recorded from 5:30 A.M. to 9:00 A.M. Similarly, at the feeding site, 64% of 515 copulations were recorded between 5:40 A.M. and 9:00 A.M.

When parties join together or when a large quantity of food is discovered, pygmy chimpanzees get very excited, and during those times, copulations occur frequently. In the evening, each chimpanzee makes its own nest, separated from others, and goes to sleep. Early morning copulations may signify greetings between a male and a female as they emerge from their respective nests after a night's sleep. Copulations during gatherings of parties probably signify the same greeting. Also, the feeding peak in the early morning may stimulate many copulations at that time.

Copulations at feeding places may be necessary in order for males and females to feed together. One example of the copulation of a young female at the feeding site clearly indicates that. The female approached a male who was holding sugarcane; after actively inviting copulation, she moved in front of him and took the sugarcane from his hand as if it were a matter of course. Scarcely any male would resist that.

The reverse order of events also occurs. First, a female faces the male and takes the sugarcane from him and, then, in the blink of an eye, presents; the male mounts as if he was caught off guard. Sometimes, a female will extend her hand toward the sugarcane and if she is refused by the male, she may also present. After mating, few males will scold and drive away a young female who tries to take the food from his hand or from in front of him.

Using this form of sexual negotiation, a female actively solicits the male and obtains food in exchange for copulation. Copulation may also be exchanged for food in another way. An old female with a pineapple was noticed and approached by the highest-ranking male, Kuma. Because he was the highest-ranking male, Kuma did not openly display begging behavior. Rather, he persistently followed this female who was avoiding him. Finally, the female presented, and Kuma copulated with her. Later, he gave up following the female around. Here, a male was currying favor by copulation; his first choice was clearly his favorite food item, the pineapple.

Sexual partners rarely groom each other immediately before or after copulation. At the feeding site, this kind of grooming was observed in only nine of 515 copulations: on seven occasions a male groomed a female, in only one case a middle-aged to old individual was involved, and in the other a young full adult or adolescent was involved. In general, grooming is regarded as a friendly behavior, but in the case of pygmy chimpanzees, sexual elements are apparently not involved.

Copulation in pygmy chimpanzees was not combined with aggressive or defensive behavior. By contrast, when Japanese monkeys copulate, males persistently attack females to the point of being sadistic, and females end up with many unhealed wounds. This kind of aggressive element is not found in the sexual behavior of pygmy chimpanzees. Instead, during the settlement of a dispute, the following scenario may take place. Upon arriving at a good food patch, tension between individuals increases, as it does when parties reunite or when members meet again in the early morning after a long separation. Copulation frequently occurs under these tension-filled conditions, suggesting that it plays a social role in preventing antagonism and facilitating peaceful coexistence between males and females. From another point of view, it may be unnecessary for a male and a female in a relaxed relationship, for example engaged in grooming, to copulate.

Frequency of Copulation

Frequency of male copulation. The chances of a male copulating depend on his age and rank. Rank is determined by the outcome of ago-

Table 18. Copulatory frequency among males at the feeding site
(January to March, 1978; January to February, 1979)

Name of individual	Number of copulations	Number of hours observed	Frequency of copulations per hour	Average frequency of copulations per hour	
Kuma	81	218.0	0.37		
Yasu	113	271.3	0.42	High-ranking male, 0.27	
Ude	21	135.3	0.16		
Hachi	45	226.1	0.20		
Yubi	41	215.5	0.19		
Kake	15	197.0	0.08		Adult male, 0.14
Ibo	74	262.6	0.28		
Kuro	6	63.0	0.10	Middle- to high-ranking male, 0.06	
Ika	4	66.8	0.06		
Hata	1	67.0	0.01		
Koshi	2	104.6	0.02		
Masu	0	35.7	0		
Hanajiro	0	17.8	0		
Mopaya	0	4.4	0		
Mon	71	268.6	0.26		
Jess	22	105.6	0.21		
Golo	11	141.5	0.08	Adolescent male, 0.13	
Futa	0	17.0	0		
Koguma	2	19.1	0.10		
Unidentified	6				
Total	515				

nistic interactions, such as supplantation of one male by another at the artificial feeding site. In adult males, the higher the rank, the higher the frequency of copulation (Table 18).

Although others do not interfere much with copulation, even if a dominant male does not interfere directly, a subordinate male may have difficulty entering into sexual interactions with a female. The spatial arrangement of individuals that were at the feeding site may explain this. The lower the rank of the males that came to the feeding place, the shorter was their stay. Most lower-ranking males took some sugarcane and hurriedly withdrew so that they could eat leisurely in a safe place, where they would not be attacked by high-ranking individuals. By contrast, many females remained at the feeding site for a long time and ate there, providing them with many opportunities to mate with the remaining high-ranking males. Even some low-ranking males stayed for a long time at the feeding site. One such individual was Ibo, who, for a low-ranking male, showed a high frequency of copulation (see Table 18). The

Table 19. Copulatory frequency among females at the feeding site (January to March, 1978; January to February, 1979)

Female group	Total number of individuals	Number of copulations observed	Range in frequency of copulation per hour per individual	Total observation hours	Average rate of copulation per hour	Range in the number of observed copulations per individual
Adolescent female	13	272	0–69	541	0.559	0–1.333
Mother with < one-year-old infant	6	23	0–18	251	0.056	0–0.222
Mother with one-year-old infant	9	26	0–14	271	0.090	0–0.151
Mother with two- to three-year-old	5	118	0–63	486	0.230	0–0.558
Mother with four- to six-year-old	5	37	0–22	444	0.085	0–0.289
Female without offspring	3	34	5–21	275	0.168	0.049–0.389
Totals and averages	41	510	0–69	2268	0.256	0–1.333

copulatory frequency in adolescent males, which is higher than in low-ranking adult males, is similarly attributable to the amount of time spent at the feeding site.

Frequency of female copulation. Females differ in copulatory frequency depending on their age and reproductive stage. The frequency of copulation was highest among young nulliparous (adolescent) females and lowest among females having infants less than one year old. Multiparous females with an offspring older than one year had a frequency between the other two (Table 19).

The strongest factor influencing female copulatory frequency is female sexual physiology. The shape and size of the female's external genitalia change with age. Until the first half of the juvenile period, the external genitalia are a slightly distorted triangular shape and very small. When the female reaches five or six years old, the size increases a little and the genitalia become a small heart-shape, but there is no swelling. At this age, females still do not engage in true copulation with intromission. Later, swelling occurs not just along the periphery of the vagina, but includes part of the anus and projects in hemispherical form in the rear. From this period on, intromission is possible.

After females give birth for the first time at 13 or 14 years old, their genitalia enlarge with age, growing in a descending direction and reaching their maximum halfway through middle-age. Some genitalia are as large as extra big bread rolls. The shape is variable and includes oval, globular, long and narrow, and nearly square. The color is usually pink, but the variation is great, from a very whitish pink to a deep reddish tinge. The size, shape, and color of the genitalia, and the manner in which folds are impressed, are fixed depending on the individual. An observer familiar with the shapes can even identify a female just by looking at her genitalia.

The external appearance of the genitalia changes periodically, as the genitalia cyclically swell and flatten. This cycle lasts an average of 46 days in captive pygmy chimpanzees (Handler, Malenky, and Badrian, 1984), and copulation occurs at the time of maximum swelling. In my field observations, almost all copulations occurred when the female's genitalia were close to or at maximum tumescence. This cycle of tumescence and detumescence is called the estrous cycle.

The pattern of estrus varies depending on the female's age and her physiological state. An adolescent female pygmy chimpanzee has an irregular estrous cycle, and maintains maximal or nearly maximal tumescence. Her external genital organ rarely deflates, but if it does, it returns to the inflated state within one to several days. Thus, these females are almost continuously in estrus and receptive to males.

The adolescent female does not get pregnant during a five to six year interval after she starts to engage in mature copulation. While in this period or condition, called "adolescent sterility," the female actively engages in copulation; she may conceive for the first time, however, between 12 and 13 years of age. If she becomes pregnant, the sexual swellings will continue at first, but then sag about two months before delivery. She will continue to copulate, however, up until one month before delivery.

For a while after giving birth, a female does not exhibit any sexual swelling. In the pygmy chimpanzee, this "postpartum anestrus" does not last long. Within less than one year after giving birth, some female pygmy chimpanzees resume their sexual cycle, regain maximum tumescence, and resume copulation. Large individual differences, however, were seen in the estrous patterns of females that resumed their sexual cycle. For example, one middle-aged female (Kame) had a full or nearly full swelling that lasted about 60 days, but another female's swelling lasted only 3 days. Nevertheless, many females displayed more than 20 days of continuous maximal tumescence.

The frequency of copulation in females may be explained by the differences in estrous patterns. Young females have the highest frequency, and infants less than one year old have the lowest frequency. The intermediate frequencies of other females may be explained by the above-mentioned differences in the pattern of their sexual cycle. If we use the time of maximal tumescence as our criterion, however, the copulatory frequency of adolescent females (0.69 times per hour) is higher than that of other classes of females (0.13–0.31 times per hour). Nulliparous adolescent females are clearly the most sexually active.

Comparison with Common Chimpanzees

Among the great apes, distinct courtship displays are found in only the two species of chimpanzees. In the gorilla, the male has a modest signal, which consists of lightly touching the female's body with the back of his hand (Nadler, 1975). A courtship display by the orangutan has not been reported.

Jane Goodall (1968) discusses six forms of courtship display in the common chimpanzee male: (1) bipedal swagger, (2) sitting hunch,

(3) branching, (4) glare, (5) beckoning, and (6) tree leaping. Forms (1), (2), (3), and (5) are also seen in the pygmy chimpanzee, but forms (4) and (6) are not. The latter two may appear threatening to the female. On the other hand, the common chimpanzee does not exhibit the "peering" behavior that includes "amae" (an infantlike request) of a young pygmy chimpanzee male. Researchers are beginning to see species differences in the attitude of the male toward the female—perhaps a difference between being tyrannical or feminist.

Toshisada Nishida (1979b) saw two new courtship displays performed by common chimpanzees at Mahale, "leaf-clipping" and "bipedal pelvic thrust." In leaf-clipping behavior, a male rips apart a leaf from a tree with his teeth, and the noise attracts a female's attention. Bipedal pelvic thrust is the straightforward performance of a copulatory movement, and it manifests frustration. Any low-ranking male may perform these two displays. Because high-ranking common chimpanzee males have a strong tendency to monopolize estrous females, low-ranking males may have devised these courtship methods to solicit females without attracting the attention of high-ranking males.

In the pygmy chimpanzee, neither these two courtship displays nor the monopolization of estrous females by high-ranking males has been observed. By contrast, I have seen an estrous female who copulated exclusively with the top-ranking male between February and March of 1978. The male (Kuma) mated with other females, but the female (Shiro), who always stayed close to this male, did not copulate with other males. The male did not exhibit any behavior that might prevent the female from interacting with other males. Thus, the top-ranking male clearly was not monopolizing this female, and, apparently, she had selected him as an exclusive mating partner.

According to Jane Goodall (1968), the male common chimpanzees of Gombe Park take the initiative in practically all copulations. There are even a few examples where a male has overpowered an unwilling female in order to copulate; in other words, "rape" does occur among common chimpanzees. Among pygmy chimpanzees, however, it is often unclear who first solicits copulation, and the female's attitude toward copulation is conspicuously positive. She presents before every copulation, and rape has not been reported.

Common chimpanzees assume various copulatory positions, but all of them are within the spectrum of the dorso-ventral position. They do not copulate ventro-ventrally, a clear point of difference with pygmy chimpanzees. The ventral position, however, is not unusual among the

Table 20. Duration of copulation among the great apes

Type of great ape	Duration of continuous copulation	Number of thrusts
Pygmy chimpanzee		
Wild	15.3 sec (1–52, N = 121)	43.8 (14–78, N = 16)
Captive	20.0 sec (Patterson, 1979)	25
Common chimpanzee		
Wild	7.0 sec (3–15, N = 45) (Tutin and McGinnis, 1981)	8.8 (3–30, N = 1,084)
Captive	8.0 sec (5–14) (Yerkes, 1939; Yerkes and Elder, 1936)	
Gorilla		
Wild	96.0 sec (30–310, N = 11) (Harcourt et al., 1981)	27.5 (6–52.5, N = 8)
Captive	52.5 sec (20–110, N = 11) (Hess, 1973)	
Orangutan		
Wild	10.8 min (3–28 min) (Galdikas, 1981)	
Captive	900.0 sec (Nadler, 1977)	

great apes. Orangutans begin copulating in a ventro-ventral position, and captive gorillas also sometimes take a ventro-ventral copulatory position. Ventro-ventral copulation, often thought to be a human characteristic, may have originated with proto-hominoid ancestors of apes and humans. For unknown reasons, only the common chimpanzee lineage lost this characteristic.

Among both species of chimpanzee, the duration of a copulatory bout, which is less than 20 seconds, is much shorter than that of an orangutan, which is around ten minutes, and that of a gorilla, which is one minute to one and a half minutes (Table 20). A pygmy chimpanzee's copulatory bout, however, is definitely longer than that of a common chimpanzee (mean, 15.3 sec vs. 7 sec). Also, a pygmy chimpanzee's bout consists of five times more thrusts per copulation than that of a common chimpanzee (43.8 vs. 8.8 thrusts per copulation). Although pygmy chimpanzees have the same basic pattern of copulatory behavior as common chimpanzees, they invest a greater amount of energy in each copulatory act.

For a long time, common chimpanzees were said to be promiscuous and, certainly for the majority, this is true. C. E. G. Tutin (1979), however, recently discovered at least two mating forms that are exceptions to the rule. The first is the so-called "consortship," in which a pair consisting of a male and a female live apart from the other members for one to several days. The second form is "possessiveness," in which a male within a party monopolizes a certain female by coercion. According to Tutin, the majority of conceptions in common chimpanzees are the products of "consortships" and "possessiveness," not "promiscuity." Pygmy chimpanzees are extremely promiscuous, and we have not observed other forms of mating. This promiscuity may relate to the greater laxity of sexual competition between male pygmy chimpanzees.

A common chimpanzee copulates more frequently than a gorilla or an orangutan (although the amount of energy invested in one copulation is small) and is said to be the most sexually active ape. Comparative copulatory frequencies measured under identical conditions are difficult to gather, but from the data collected so far, pygmy chimpanzees appear to be much more sexually active than common chimpanzees. Under both natural conditions and provisioning, the pygmy chimpanzees of Wamba copulate much more frequently than common chimpanzees (Table 21).

Captive-reared chimpanzees engage in sexual behavior at abnormally high frequencies. The lack of need to perform life-sustaining activities such as foraging, feeding, and nest-building may account for this pattern. For example, 107 instances of copulation were recorded during 100 hours of observations of the American A.R.L. colony of common chimpanzees, or 1.07 copulations per hour. Even that value, however, does not reach the copulatory frequency, 1.56 copulations per hour, of provisioned pygmy chimpanzees at Wamba. The size of the A.R.L. colony fluctuated periodically between 25 and 45 individuals, while the party size at Wamba was from 10 to 40 individuals. At times, they were almost the same size, or the Wamba party was smaller. Thus, the observed difference in copulation frequency may arise from the pygmy chimpanzee's abnormally frequent copulation.

Common chimpanzees, like pygmy chimpanzees, copulate during the female's estrous period. According to Jane Goodall and C. E. G. Tutin, most of the copulations recorded at Gombe occurred when the female's sexual skin was close to or at maximum tumescence. The number of copulations per estrous female per hour varied depending on the individual. At Gombe, the range of rates was from 0.31 to 0.99 copulations per hour for common chimpanzees; at Wamba, it was from 0 to 1.33 copulations per hour for pygmy chimpanzees. Therefore, we cannot say that there is a difference in female copulation rate between the species at

Table 21. Copulatory frequency among the great apes

Great ape group and field site	Observation period in months	Hours observed	Number of copulations recorded	Investigator	Number of copulations per hour of observations
PYGMY CHIMPANZEES					
Wild					
Wamba	12	700	219		0.313
Provisioned					
Wamba	3.5	330	515	Kano, 10/75 to 2/79	1.561
COMMON CHIMPANZEES					
Wild					
Budongo	5.7	300	4	(Reynolds and Reynolds, 1965)	0.013
Gombe	24		32	(Goodall, 1965)	
Mahale	22		14	(Nishida, 1968)	
Budongo	3.9	360	19	(Sugiyama, 1969)	0.053
Bossou	6		35[a]	(Sugiyama and Koman, 1979)	
Provisioned					
Gombe	12		213	(Goodall, 1968)	
Gombe	15	1,200	1,137[b]	(Tutin, 1979)	0.95
Mahale	35		383	(Nishida, 1979a)	
Captive					
Delta Regional Research Center	7	1,200	341	(Tutin and McGrew, 1973a)	0.284

ARL colony	1	100	107	(Kollar, Beckwith, and Edgerton, 1968)	1.07
University of Oklahoma		20,564 days[c]	666	(Lemmon and Allen, 1978)	
Delta Regional Research Center	5	741	213	(Tutin and McGrew, 1973b)	0.287
GORILLA					
Wild					
Kabara	19	466	2	(Schaller, 1965)	0.004
Virunga Mts.					
Mt. Visoke	1.5	75	0	(Elliot, 1976)	0.000
Mt. Virunga	30	2,000	98	(Harcourt and Stewart, 1978)	0.049
Kahuzi	6	350	2	(Yamagiwa, Kano)	0.006
ORANGUTAN					
Partially Wild					
Northern Sumatra	36	5,000	58	(Rijksen, 1978)	
Kalimantan	41		37	(Galdikas, 1979)	0.007
Wild					
Sabah (Borneo)	16	1,200	7	(MacKinnon, 1979)	0.006
Northern Sumatra	7	300	3	(MacKinnon, 1979)	0.01

[a]Including immature individuals   [b]Observations of only estrous females   [c]Total number of days by focal animal sampling

Gombe and Wamba. Other factors account for the differences between common and pygmy chimpanzees in the overall copulatory frequencies of groups.

First, there are differences in the level of sexual activity of adolescent females. Goodall (1968) discovered that adolescent female common chimpanzees, whose sexual swelling is not attractive to males, do not copulate very often. By contrast, adolescent female pygmy chimpanzees are in the age class that copulates most frequently.

Second, there are great differences in the length of sexual inactivity before and after parturition. Among common chimpanzees, a female does not resume her estrous cycle for 30 months (primiparous female) to 48 months (multiparous female) postpartum (Goodall, 1983), and during this period, she does not copulate. In contrast, pygmy chimpanzees usually resume estrus and copulation much earlier.

Third, there are differences in the duration of maximum swelling in a normal estrous cycle (coinciding with the menstrual cycle). In wild common chimpanzees, the duration is 9.6 days (25% of a cycle of 37.2 days) (Tutin, 1979). In pygmy chimpanzees, the period of maximal swelling usually lasts for more than 20 days (more than 43% of a cycle of 46 days), or twice as long as for common chimpanzees. The long estrous period of a female pygmy chimpanzee relative to a common chimpanzee, no matter what her physiological stage, certainly results in a higher copulatory frequency.

The menstrual cycle of an adolescent female common chimpanzee, 42.6 days on average, is longer than that of an adult (Young and Yerkes, 1943). Furthermore, the cycle of sexual swellings is very irregular, and the period of maximal swelling is longer than that of an adult. The adolescent female pygmy chimpanzee is almost always in an estrous condition, and should not be used for comparisons. The *adult* female pygmy chimpanzee, however, has a sexual physiology similar to that of an *adolescent* female common chimpanzee. First, the length of the menstrual cycle of an adult female pygmy chimpanzee, 46 days (Handler, Malenky, and Badrian, 1984), is closer to that of an adolescent female common chimpanzee than to that of an adult female common chimpanzee. Second, the pattern of sexual swelling of the adult female pygmy chimpanzee is irregular and long, resembling that of an adolescent female common chimpanzee. Perhaps the female pygmy chimpanzee is neotenous in its sexual physiology.

That the birth interval in the common chimpanzee, on average between 5 and 5.5 years (Goodall, 1983), does not differ from the pygmy chimpanzee's shows that the high copulatory frequency of the pygmy chimpanzee does not accurately reflect the rate of reproduction. Goodall (1968) reports that, "After most copulations observed from close quarters, the males were seen with ejaculate adhering to their penis." On the contrary, at Wamba, we could not find evidence of semen in the majority of copulations involving pygmy chimpanzees. Perhaps a low rate of ejaculation accounts for their low conception rate.

In the common chimpanzee, ovulation occurs shortly after the middle of the period of maximal tumescence, and copulation occurs most frequently during this period when the likelihood of conception is greatest. In the pygmy chimpanzee also, ovulation probably occurs at a time when the sexual skin is maximally tumescent. Because the period of maximum swelling is very long, however, the probability that copulation will coincide with ovulation and result in pregnancy must be lower.

Following postpartum amenorrhea, a female common chimpanzee resumes menstruating and will conceive again in an average of 3.6 cycles, about 4½ months (Tutin and McGinnis, 1981). In contrast, a female pygmy chimpanzee resumes her cycle of sexual swelling earlier than a common chimpanzee, but she does not conceive for at least three years after resuming copulation. This prolonged absence of conception is difficult to explain in terms of only the low ejaculatory rate and the longer period of maximal tumescence. A female pygmy chimpanzee may be sterile during this period even though she is in estrus.

The relationship between postpartum amenorrhea and breast-feeding the newborn infant is well known. Because both species of chimpanzee nurse their offspring up until the age of four years, the female may be sterile (not ovulating) during that period. A female common chimpanzee does not have estrus during this period, but a pygmy chimpanzee resumes her cycle much earlier, when she is still sterile. I suspect that "reproduction" and "estrus" are decoupled in the pygmy chimpanzee.

Pygmy chimpanzees and common chimpanzees are very similar in their patterns of courtship and copulatory behavior and in their sexual physiology, based on the cycle of tumescence and detumescence of the female genitalia. These similarities result from the close affinity of the two species. In the gorilla and the orangutan, sexual swelling is barely or not at all visible.

Pygmy chimpanzees differ from common chimpanzees in that pygmy chimpanzees copulate excessively. Because most of the copulation does not result in conception, the activity seems wasteful from the point of view of reproduction. Although copulation is essential for reproduction, from the perspective of day-to-day maintenance, it merely constitutes a loss of energy. Also, because the engaged pairs are practically defenseless during copulation, it is normally a time of unsurpassed danger. Consequently, many animals, including the common chimpanzee, have shortened the duration of copulation when possible and, perhaps, natural selection has worked to reduce the failure rate of conceptions.

In contrast to the common chimpanzee, the pygmy chimpanzee may have undergone selection for a high frequency of copulation. Female pygmy chimpanzees are much more sexually receptive than female common chimpanzees, and this has provided the motive force behind the vigorous sexual behavior of pygmy chimpanzees. Copulation is extremely inefficient, however, and very few pregnancies result. Nevertheless, the distinctive characteristic of pygmy chimpanzee society—that males and females are always together—may be maintained by this frequent copulatory behavior. Here, copulatory behavior goes beyond reproduction. In pygmy chimpanzee society, the primary role of copulatory behavior is undoubtedly to enable male-female coexistence, not to conceive offspring. In this case, we ought to recognize copulation primarily as a social behavior, with "reproduction" as a secondary function.

CHAPTER SEVEN

Social Behavior and Social Relationships

Kin Relationships

Social behavior is not directly tied to individual survival and reproduction. But because relationships between individuals are determined through social behavior in higher animals that live in groups, social behavior provides the most important framework for establishing the species-specific patterns of individual existence, reproduction, and care of offspring.

A social group is a gathering of heterogenous individuals, each with his or her own role to play. By carrying out their respective roles, individuals work for their own mutual coexistence and maintain the group. The ratio of males to females in a pygmy chimpanzee group (unit group) is 1:1, whereas it is from 1:1.2 to 1:2.2 in the common chimpanzee. Also, the party size in pygmy chimpanzees is larger. These two differences hint at different mechanisms for individual coexistence in the two species of chimpanzee.

In this chapter, we describe how the coexistence of individuals within a pygmy chimpanzee group is achieved through their social relationships. The components of a unit group are roughly divided by sex, into male and female classes, and by age, into infant, juvenile, adolescent, and adult stages. The unit group may be further classified into kin and non-kin members. The central kin relationship is the mother and her offspring; relationships with and among siblings (brothers and sisters) are secondary. A father-son relationship ought to exist within the group, but the promiscuous mating system makes the pertinent research impossible at present.

Mother-offspring relationships. The infant clings to its mother's belly for a long time after birth, so that initially the researcher has difficulty even determining the infant's sex. Moreover, we cannot be certain when an infant first separates from its mother's belly because research at Wamba is intermittent. We estimate, however, that separation occurs after the infant is three months old.

At about seven to eight months of age, the infant will occasionally cling to and be transported on the mother's back, but it will usually be carried ventrally up until two years of age. In crisis situations, the mother will carry even a four-year-old ventrally.

At two years of age, a juvenile is regularly carried on its mother's back. When the mother stands still or moves at a very slow pace, the juvenile sits up straight in the middle of the mother's back or sits turned sideways. When the mother begins to walk briskly, however, the juvenile quickly takes a prostrate position and clings to her back. When the offspring reaches three years of age, the juvenile will be carried when the mother is moving on the ground, but when the mother begins to ascend a tree, the juvenile will normally dismount quickly. In this situation, the juvenile will travel in front of its mother more often than behind her.

The infant and juvenile leave the mother mainly at resting and feeding times. When the infant reaches six months of age, it starts to move around the periphery of the mother's body. Especially when in the trees, it will leave the mother's body to dangle from a small branch, and will start to take tottering steps. But the infant's movement is clearly limited to a range within reach of its mother. If the infant attempts to go farther than that, the mother will bar its way with her hand and will resume carrying it.

As the months and years go by, the distance from the mother and the time spent away from her gradually increase. Nevertheless, the mother often carries her offspring during travel until it is at least three or four years old. The signal initiating this kind of transportation is the mother's vocalization. Then, after walking a short distance, up to 6 m, she will stand with one foot slightly lifted, the sole facing toward the rear, in a stationary walking position. There she will stand, waiting for the juvenile to run after and jump onto her back. In the trees, when it is time to move, she will assume an ascending or descending posture and wait. When her offspring is close, she may extend her hand toward him.

During rest periods, the mother often becomes her infant's playmate. The play motions between mother and infant are slow-moving and gentle. She may tickle with her fingers, grab with her hands, and playbite. When they are in the day nest or on the ground, the mother may lie on her back and raise the infant up with her hands and feet in playing "airplane." After the infant enters the juvenile stage, mother-offspring play is rare. The mother begins to groom her infant after it is five to six months old, and the amount of grooming increases thereafter.

According to L. C. Miller and R. D. Nadler (1981), the time spent breast-feeding in common chimpanzees decreases by about one-third between birth and six months. This trend is the same as in the orangutan, suggesting that the development of nutritional requirements among infant great apes is practically identical. Although nursing females are common among the pygmy chimpanzees of Wamba, there are no good observations of nursing and other mother-infant activities because observing them at close range is difficult. Nursing probably lasts until the offspring is about four years old.

Sometimes juvenile common chimpanzees have a "tantrum," which Goodall calls a "temper tantrum." During this tantrum, a frustrated juvenile emits a plaintive cry, "ho-ho-ho," that crescendoes to intense screaming, "gya-gya," while the juvenile rolls on the ground holding its head. The tantrum is easily dispelled by embracing the mother and clinging to her nipple. Although the causes of these tantrums, such as the mother not sharing her food, are variable, the effectiveness of the mother's nipple is striking. We have rarely seen a pygmy chimpanzee mother at Wamba actively reject her offspring for the purpose of weaning.

At five or six months of age, pygmy chimpanzees begin to take an interest in their external surroundings. At about this time, they begin to put solid food in their mouth. They will extend a hand toward food in their mother's mouth or hand, and they will suck on abandoned food scraps and try to eat them. At about this time, they also begin to show begging behavior and will frequently solicit food from their mothers until the end of the juvenile period. When food, such as sugarcane, has a skin or covering that is difficult for the juvenile's teeth to tear, the juvenile may take food that has already been stripped by its mother.

The most prevalent form of juvenile begging behavior is touching the mother's mouth. The juvenile, while sitting on its mother's lap or clinging to her back, touches her mouth as soon as she puts the desired food in her mouth. The mother moves her head as if annoyed and looks the other way, but she does not brush aside the juvenile's hand. Finally, she gives up the food from her mouth. Occasionally, a mother will strip the skin from a piece of sugarcane, break off a piece, and hold it in her mouth, without chewing, until her juvenile offspring takes it.

Among juvenile and infant common chimpanzees, putting the mouth near the mother's mouth is the most generally observed begging-like be-

An infant male, Tawashi, persists in begging for food from E-group's alpha male, Kuma.

Table 22. Food-sharing relationships

Sharers and recipients	Number of cases of food-sharing
Between non-related individuals	
male → male	25
male → female	126
female → male	65
female → female	81
male/female → juvenile/infant	70
juvenile/infant → male/female	5
juvenile/infant → juvenile/infant	2
Total	374
Between related individuals	
mother → offspring	127
offspring → mother	6
sibling → sibling	2
Total	135

NOTE: Based on data collected by Kuroda, 1984, on about 509 cases of food sharing at the feeding site. There were many cases of nonrelated males sharing with nonrelated females, but there were half as many cases of nonrelated females sharing with non-related males. In common chimpanzees, no one has yet seen a female share with a male. The total number of cases of food-sharing among related individuals is one-fourth of the total number of food-sharing incidents.

havior. Among pygmy chimpanzees, only one two-year-old male called Haku did this often, and the behavior was seldom seen in other individuals. Instead, a kind of food-sharing occurs frequently in which a juvenile approaches and snatches food from its mother or takes food directly from her mouth. The mother certainly does not dole out the food, but she lets her offspring pull and bite at it (Table 22).

Male juvenile chimpanzees vigorously pursue sexual interactions with adult females. Such "immature copulations," however, are very unusual between a mother and son. During observations of 137 mother-son units, only five cases were recorded. In the pygmy chimpanzee, the avoid-

The infant Tawashi begs for food from his mother, Kame.

An example of immature copulation: a juvenile male gives a sexual display while touching an adult female.

The female permits the young male to copulate, dorso-ventrally.

ance of mother-son mating (incest avoidance, in human terms) is established at an early age.

Upon reaching the late juvenile stage (5–6 years old), an individual begins to spend less time with his mother. Sometimes, the juvenile at this stage will resume begging from his mother and grooming her, but he will play less with her and more with older individuals and peers. The juvenile is never much more than 20 to 30 m from his mother. When his mother gets up to travel, he seems to appear from out of nowhere and rushes toward her. Even though separated from his mother, the juvenile is usually attentive to her behavior.

As females approach adolescence, they become less social than males. They occupy the periphery of the group and often sit alone in a tree. This may be preparation for leaving the group and, occasionally, emigration occurs suddenly after a female has entered the adolescent stage. Because adequate data are available for E-group only, we do not yet know how a female who emigrates from her natal group finds her way to another group, nor do we know into which group she settles. Nevertheless, none of the adolescent females who had emigrated from E-group returned to their mother's (natal) group even temporarily. At some point during the daughter's adolescence, the mother-daughter relationship is completely severed.

On the other hand, the mother-son relationship continues for a very long time. Throughout adolescence and even after entering adulthood, the son almost always forages with his mother, and grooming occurs most often in this relationship. Occasionally, an adult son will even display "begging behavior" to his mother. The converse—the mother showing "begging behavior" and the son sharing food with his mother—is extremely rare.

An old adolescent male, Mon (left), begs for food from his mother, Kame (middle). Mon's elder brother, Ibo (right), is eating sugarcane that Kame had shared with him.

A young adult male, Ibo, begs for food from his mother, Kame.

An adult female pygmy chimpanzee with an infant begs for food from another adult female with an infant. By contrast, food sharing is rarely seen between female common chimpanzees.

Protection of offspring by the mother is one of the typical maternal behaviors of pygmy chimpanzees. For example, the female Shiro was once observed protecting her six-year-old daughter Shiroko from an attack by the higher-ranking female Haru, who was known for her violent behavior and who would often bully Shiro. In this case, when Shiroko stole a piece of sugarcane from in front of Haru, Haru pounced on Shiroko, pinned her down, and bit her. Shiro, who had been eating only 3 m away, instantly rushed over and squeezed herself between them. Then, Shiro stood Haru up bipedally, soothingly embraced Haru's shoulders with both hands, and gently pushed her until they were 2 m away from Shiroko. While being pushed, Haru kept glaring at Shiroko, who was screaming but did not show any further aggression toward her.

Maternal care behavior decreases as the juvenile matures, but the mother occasionally gives a warning vocalization and intervenes by attacking when her adolescent or adult son is attacked. By contrast, a son rarely steps forth on behalf of his mother. Because females, especially those beyond middle-age, are rarely attacked, perhaps there are few opportunities for a son to offer protection. The following, however, is a rare example.

A young adult male, Ten, attacked and bit an old female, Kame. Kame, who was being held down, screamed, and her son Mon, who had been feeding about 20 m away, rushed toward Ten and leapt onto his back. In succession, Mon's older brother Ibo and Ten's mother Sen sprang up and piled on top of the others. All we could see was a black flurry. Mon was the first to flee into the bushes, followed by Ten, Sen, and Ibo. Kame, who brought up the rear, noticed that she had shaken her infant off and, confused, turned back. At that time, the dominant-subordinate relationships between the males were, in descending order of rank, Ten, Ibo, and Mon.

In general, female pygmy chimpanzees are very indulgent mothers. They do not punish or threaten their offspring at all. Haru, however, was an exceptional mother and was on occasion seen wresting sugarcane from her juvenile son Haruo. The rest of the mothers were never seen committing this kind of misdeed.

A male pygmy chimpanzee of Wamba spends a great deal of time with his mother. When a son enters adulthood, he never displays dominant behavior toward his mother. The mother indulges her son and does not show any fear of him. Why does pygmy chimpanzee society continue to produce such mother-centered sons? What are the selective advantages as interpreted by sociobiological theory? These are questions for future research.

Sibling relationships. Because the birth interval of pygmy chimpanzees is more than five years, a mother rarely has two or more offspring belonging to the same age class at the same time. Consecutive siblings follow the order infant, juvenile, adolescent, and young adult. But when a mother is successful in giving birth within a short interval, the next youngest child is still nursing and at an age when normally he would be riding on his mother's back. By virtue of the new child being born, this older infant suddenly loses the benefits and protection of its mother. The behavior of the older child during this period, however, has not yet been well studied.

The way a mother treats an older child seems to vary with the mother's personality. For example, a female named Fuji usually carried her newborn infant ventrally and the older child dorsally. She continued to transport both of them for two years. When a third was born, she behaved in the same way. The oldest child showed great interest in, but no conspicuous jealousy toward, his sibling. He established an amicable relationship with his infant sibling, grooming and playing with it.

By contrast, a younger child does act jealous of an older sibling. In another example, Shiro's two-year-old offspring Haku pushed his mother away with his slender hand when she would approach his six-year-old sister Shiroko.

An example of a strong attachment toward a younger sibling involved Tawashi (the adolescent third son of Kame) and his one-year-old little sister Kameko. Tawashi often approached his mother, peered into her face, and after looking awhile carried Kameko and took her for a walk. Once, he made a nest 10 m from Kame and, lying on his back, played with Kameko on top of his stomach. He tickled her, held her up by the arms, embraced, and kissed her (open-mouth kiss), pressing his large open mouth everywhere on her body. The infant tired and began to whimper, but Tawashi continued until their mother heard and came running.

Tawashi's behavior was also impressive when Kameko, who was underdeveloped, died. When we found Kameko dead, her small body was being held and carried around by Tawashi. He carried his little sister's body with all four limbs hanging down lifelessly; one of his arms pressed her against his chest; and he walked slowly in the tree apparently in deep thought. The mother, Kame, walked in front, clearly preoccupied and dispirited. The second son, Mon, followed several meters behind Tawashi. Several times Mon tried to touch Kameko, but Tawashi quickened his pace or closely embraced the corpse with both arms to conceal her and turned his back to avoid interference by his older brother.

This unfortunate family separated from the other members of the group and foraged on their own. After several minutes, they descended to the ground and I lost sight of them. In the evening, when they ascended the trees to make their nests, the older brother was still holding

Kame sitting by her dead infant, Kameko.

his dead sister. He lay the corpse in the fork of a branch and, after adjusting the body several times to stabilize it, he moved to the end of the branch and made his own nest.

The following morning, the first to leave his nest and approach the corpse was Tawashi. He lightly touched the corpse and, after grooming it a little, he began to wrestle with his older brother Mon who had approached. Their mother, Kame came later. She lingered near the corpse and stared at it. As the sun rose, flies started to swarm around the corpse. Several times Kame grabbed quickly at the air with her hand as she shooed them. This behavior continued for several minutes after the family was joined by other members of the southern subgroup, making a large commotion. Then Kame picked up the corpse in her arms and concealed it.

For a while, Kame's family seemed, in general, to live separate from the others. Then the family left together and did not come back to the feeding site. Six days later, when we found Kame, she did not have the corpse. Kame was depressed for a long time, however, and had few social relationships with others.

Among primates, it is not unusual for a mother whose infant has died to carry the corpse around for several days. According to Western researchers, the mother's attachment to her dead infant can be explained physiologically as well as socially. For example, J. J. McKenna (1982) reported that a mother langur carried her dead infant over a five-day period because her breasts were swollen and needed to be emptied to relieve the discomfort. Such a physiological explanation, however, does not account for an older brother carrying the corpse. In chimpanzees, we may have to admit that feelings exist that are similar to those of human beings. Kame's family usually had the best rate of attendance at the feeding site; they liked sugarcane and were sociable. Their avoidance of other individuals for several days and their lack of attendance at the feeding site were atypical for this family. Nevertheless, we may be anthropomorphizing this situation because we can imagine ourselves being overwhelmed by an unexpected misfortune and losing our appetite from grief.

Tawashi's "babysitting" and carrying the corpse of the infant are examples of "allomothering" behavior. Allomothering behavior by adolescent females has been reported for the common chimpanzee, but this behavior by a male toward his younger sibling has not yet been reported. In the pygmy chimpanzee, such behavior may not be rare. A young male, Koshi, who was the presumed oldest son of an E_2 female named Tare, would often carry his two-year-old male sibling Tareo on his back.

Adolescent and older pygmy chimpanzee males are very tolerant of their infant or juvenile siblings and often groom them. After reaching adolescence, only the males have sibling associations because females leave the natal group. Usually males travel together in the same party and often engage in mutual grooming, though the frequency decreases with time. Occasionally these mature male siblings also engage in "play."

Males begin to have dominant-subordinate relationships in adolescence. As far as we know, the older brother, without exception, is the dominant individual. The older brother may approach his younger brother in the act of copulation and slap him, grab his hips, drag him, and bring him down. Or, the older brother may tear sugarcane away from his younger brother and seize it for himself. Such sudden, aggressive body contacts between unrelated males do not often occur. In response to this seemingly unwarranted treatment by an older brother, the younger brother usually withdraws in silence, although occasionally he may counterattack while screaming protests. Aggression coming from the subordinate's side is also an unusual phenomenon among unrelated males. Sometimes, fights between male siblings shift to play wrestling, which may be initiated by either one of them by making a "play face" (opening the mouth halfway and barely showing the teeth) and uttering a "play pant." While both males are still young (adolescent or young adult), sibling rivalry can be unconventional. Male siblings seem to have a mutual tolerance for each other that probably modifies any aggressive behavior between them.

In common chimpanzees, alliances between male siblings are known to greatly influence their rank. A male named Figan at Gombe Park finally succeeded in rising to the highest rank (alpha) with the assistance of his older brother Faben (Riss and Goodall, 1977). Faben was a victim of "infantile paralysis" (polio) that had spread throughout the Kigoma region. Despite being paralyzed in both arms, he acquired high rank, and was able to maintain it because of the support of his higher-ranking younger brother.

In pygmy chimpanzees, there are no clear-cut examples of this sort of alliance of blood relatives. For example, Kame has three sons: the eldest, Ibo, is in the prime of life; the second son, Mon, is in the spring of his youth and at his strongest; and the third son, Tawashi, is just entering adulthood. If there were intimate mutual support between the three brothers, they might be able to control most of the southern subgroup to which they belong. They do not, however, and even with three of them, the young Ten could always chase after them. Moreover, the aged male Kake had an uncommon dislike of Tawashi. Kake waited for Tawashi to enter the feeding site, attacked him, and drove him away. Nevertheless, Tawashi's two older brothers, Ibo and Mon, who were about equal in rank to Kake, were quite indifferent to the distress of their younger brother. Instead, Tawashi always fled to find the support of his mother, Kame.

The bond or intimacy between postadolescent siblings is primarily mediated by the mother and probably is not very strong. One subject of future interest is how the relationships between siblings change after their mother's death.

Relationships Between Adult Males

One important element characterizing the relationship between adult males is aggressive interactions. Adult males have the highest fre-

Table 23. Agonistic interactions

Individuals who showed aggressive or dominant behaviors	Individuals who showed defensive or submissive behaviors						
	Adult male	Adult female	Adolescent male	Adolescent female	Juvenile male	Juvenile female	Total
Adult male	163	82	14	11	3	1	274
Adolescent male	1	16	3	2	2	1	25
Adult female	5	9	6	1	2		23
Adolescent female		1				1	2
Juvenile male		1					1
Juvenile female							0
Total	169	109	23	14	7	3	

NOTE: January–February 1978, under provisioned conditions. Includes 23 examples of mutual aggression or locomotor "play" activity in which dominant-subordinate behavior was not obvious. For example, one pygmy chimpanzee approached another, and the latter, without acting subordinate, moved out of the way. Furthermore, in the agonistic interactions listed, the total number of individuals involved was 632. Of that total, 425 were adult males; 135 were adolescent males; 46 were adult females; 16 were adolescent females; 8 were juvenile males, and 3 were juvenile females.

quency of aggressive interactions, relative to associations between members of other age-sex classes (Table 23).

The term "aggression" has been defined as behavior in which one or several individuals injure and threaten or try to injure another (McKenna, 1983). Aggressive behavior is inevitably accompanied by a reaction. The reaction may be aggressive, or it may be submissive. We also use the term "agonistic" behavior for behaviors exhibited by both aggressor and victim.

The aggressive behavior pattern of pygmy chimpanzees abounds with variety, from violence, including physical contacts such as biting, hitting, kicking, slapping, grabbing, dragging, brushing aside, pinning down, and shoving aside, to glaring, bluff charging (the appearance of charging), charging, and chasing. A pygmy chimpanzee may also approach another with exaggerated gestures, wave his arms around the other's head, and leap over the other's body, threatening body contact. Aggressive behavior is accompanied by a vocalization that sounds like "kat-kat."

Prostration, grimacing, flight, avoidance, extending one's hands, and touching the other's body are classes of submissive behaviors in response to aggression. Three kinds of shrieks are emitted by victims. They are, in order of increasing intensity, "gyaa-gyaa," "kii-kii," and "ket-ket."

Low-level aggressive interactions, such as approach-avoidance or

Mild aggression between males.

A typical aggressive interaction between male pygmy chimpanzees.

Wrestling or play-fighting, with both males in a bipedal stance.

A male grimaces while mounting a higher-ranking male.

bluff charge-flight, occur very frequently between males and seldom involve other patterns of behavior. Furthermore, the aggressor and victim
are constant; that is, the victim and aggressor do not switch roles.

Aggressive interactions mainly occur at feeding time and, most
often, when individuals gather in the trees to feed on concentrated foods
such as *Dialium* and *batofe*. This suggests that proximity between individuals may induce aggressive behavior.

Successive mounting or rump-rubbing often occur during aggressive interactions. Mounting is riding horseback on another, taking the
position used in dorsal copulation. In rump-rubbing, two individuals in
a "presenting" posture direct their rear ends toward each other and press
their buttocks against each other. Sometimes the positions are assumed
and held motionless, but usually they are accompanied by slight rhythmical thrusting movements. Frequently, both actors have erect penises,
but insertion into the anus during mounting has not yet been verified.

In an aggressive interaction, the attacker may spring on a male who,
cut off from escape, is groveling and screaming, and the attacker will
mount or rump-rub the victim. Or, the attacker may confront his victim,
suddenly facing the victim's buttocks and demanding mounting or rump-
rubbing. The order of these events is not fixed.

Immediately after arrival at the feeding place, a male may make a gesture similar to a sexual display to another male, and then, either mounting or rump-rubbing may occur. Mounting and rump-rubbing may have the same function because the social context in which they occur is practically the same. These actions put an end to aggressive interactions, preventing anything injurious from happening, and have the effect of "ap-

Two males rump-rubbing.

A male, who has come for sugarcane, is watched closely by a higher-ranking male, at far right.

The higher-ranking male threatens the lower-ranking male.

The lower-ranking male appeases the higher-ranking male by allowing him to mount.

peasement" or "pacification." The dominant male presents a guarantee of safety to the subordinate.

Among the behaviors similar to aggression is the charging display, which male common chimpanzees often display. The charging display includes a variety of aggressive behavioral elements, such as piloerection, violently running around on the ground, wildly leaping from branch to branch, furiously striking the ground and tree buttresses with feet and hands, and hurling pieces of wood and stones. Usually, charging displays are accompanied by a characteristic peculiar vocalization, "hooting."

A pygmy chimpanzee's charging display, in contrast to that of the common chimpanzee, is simply running while dragging a 50- to 150-cm long branch or shrub. The male vigorously stamps his feet while running 20 to 30 m, and then, all at once, he flings the branch. The foot stomping and the scraping of the branch sound almost like a startled elephant rushing through the forest. Occasionally, a pygmy chimpanzee will beat on a tree buttress or may vocalize, but most of the time, he is silent and runs in a straight or gently curved line. The pattern of branch-dragging varies little among individuals.

We do not understand many points about the social meaning of the pygmy chimpanzee charging display, or branch-dragging behavior. Branch-dragging, which frequently occurs right after arrival at the feeding site, is often directed at other individuals. For example, a male may head straight for one individual and run over him, covering the victim's body with the branch that is being dragged. Then, the male who began branch-dragging may shriek and run away. Branch-dragging behavior, therefore, seems to include some elements of threat, but does not appear to cause any reaction by group members. Branch-dragging behavior occurs in high frequency after feeding, after napping, and just prior to beginning to forage, and at these times, the effect may be to urge other members to begin foraging.

Dominant-subordinate behaviors or hostile behaviors prevail when males approach each other. These behaviors are the most important factor controlling relationships between unrelated pygmy chimpanzee males, as they are in many other primate species that form multi-male groups. Even after ten years of provisioning, however, we do not understand how rank is established among males.

We find that although dominant-subordinate interactions randomly occur in certain associations, they often do not occur in others. One reason may be that some males rarely associate with other males of the same party, while some males are characterized by frequent social interactions.

Moreover, a certain male may often confront another male with aggressive behavior, but other males hardly show any dominant behavior even though they are always together. In short, some males are very aggressive, and others show little dominant behavior.

The breadth of variation in personality among pygmy chimpanzees is so great that a simple graphical representation of the dominant-subordinate relationships between individuals cannot be drawn. There is evidence, however, that pygmy chimpanzees do not have a strictly linear rank order. Since 1977, when individual identification was almost complete at Wamba, five adult males, with one named Kuma at the top, formed a high-ranking group. All five belonged to the northern subgroup, and moved south as a unit. When these northern males merged with the southern subgroup, the unit of five males from the northern subgroup ranked higher than the males of the original, southern subgroup. But when the fourth- or the fifth-ranking male independently joined the southern subgroup, he assumed a submissive attitude toward a male in the southern subgroup. Their high rank had been maintained by the northern group's "male alliance."

In the southern subgroup at this time, there were three robust adult males who always accompanied each other when foraging. At the feeding site, if they were feeding together, dominant-subordinate interactions hardly ever occurred among them. These three formed an alliance, but it did not raise the rank of its members. Usually, it was the lowest ranked group in the southern subgroup. We often called this phenomenon "damesan" or "dametrio," the "powerless three" in Japanese.

The term "alliance" (first used to describe a common chimpanzee behavior) may or may not apply to male pygmy chimpanzees. Nevertheless, in the northern and southern subgroups of Wamba's E-group, there were eight males that were intimately associated, as of 1982. Twelve other adult males were in E-group at the time. On the basis of their home range, eight of these 12 individuals belonged to the northern subgroup and four belonged to the southern subgroup. None of these males belonged to either of the above-mentioned "alliances," nor did they form especially intimate relationships with other males.

Two of these "non-allied" males were old and two others were severely physically handicapped; they were males who had no interest in social activities or who had limited social ability. The other eight were relatively young and were accompanied by females who appeared to be their mothers; these males, rather than participating in male alliances, chose to live with their mothers. The eight males in the "alliance" were in the prime of life, and from the beginning of provisioning, we did not find any females who seemed to be mothers of these eight males.

Here again, we see the influence of the mother-son relationship on social structure. Continued intimacy in the mother-son relationship is apparently an obstacle to a strong alliance of adult males united by the "male bond." The mother can influence the male's class or rank in an even more direct way. Four cases have been reported of young males who rose in rank within a short period of time. The common characteristic of all these young males was that their mothers were middle-aged to old females respected by other individuals.

Here I consider the case of Ibo. According to Kitamura's research, Ibo's rank rose in 1980 to the extent that he showed deference only to the top-ranking Kuma. At that time, Ibo fought with the second-ranking Yasu for a long time at the feeding site, and their dashing young warrior manner is still talked about by the sugarcane carriers. Finally it was a draw, Ibo's treat being that the fight did not continue for longer. The following year when I came to Wamba, Ibo had returned to being the neighborhood urchin, quietly visiting the feeding site from the shadow of the prop roots of a *bosenge* tree on the outskirts of the feeding site. We can think of only one cause for this. His mother, Kame, was old. But she was an unusual female who was almost always in estrus, and when she was in estrus, she seemed to be very attractive to both females and males. When she arrived at the feeding place, Kame was solicited constantly for genital rubbing or copulation by other females or by males. Even high-ranking males respected her, not to mention females and young males. When I re-encountered her after a while, she was embracing a small infant. Her sexual skin was deflated, perhaps explaining why she was no longer popular. Kame had become retiring while caring for her infant, and she would even flee when followed by a middle-ranking male. Perhaps Ibo had slipped from his high rank because his mother had fallen in popularity.

Following the change, another female, named Sen, took over Kame's former position. Her son Ten began to make his influence felt, despite his youth, and two years later, he was still a central figure in the southern subgroup. Ten, who is young and still occasionally displays "childish" behavior, is now firmly in position as the alpha male.

Toward the end of my research in 1982, Koguma, the son of Aki, a powerful female in the northern subgroup, had begun to enter adulthood. One day, he suddenly aggressively charged the second-ranking male, Ude. While branch-dragging and screaming, Koguma thrust violently and forcefully at Ude and ran past very close to his body, to which Ude responded by leaping up and slapping Koguma. The top-ranking Kuma intervened, mounted Ude, and then rump-rubbed, calming Ude's agitation.

After a while, however, Koguma charged again. Ude sprang up and chased Koguma, but Koguma stood his ground at the outskirts of the feeding site and counterattacked. They flew about between shrubs and bushes, and exchanged violent blows; then, Koguma dived into the bushes, and Ude pursued him. Aki, with a screaming infant clasping her belly, rushed toward Ude. Ude fled as Aki, with the loud vocalizations of other females behind her, chased him into the bushes of the opponent's side.

Koguma's charges at Ude continued, and in the span of nine minutes, he attacked 12 times. Every time that Ude rose to retaliate against Koguma, Aki would divert him by vocalizing loudly, and running toward him. After Koguma's seventh charge, Ude became silent and avoided Koguma. Finally, I could just hear the sound of Koguma's branch-dragging from deep in the bushes, and Ude fled flying into the nearest tree.

Since that day, Ude has seemed unsettled. Every time Koguma has rushed at him, Ude has fled, taking little steps, or he has presented, as if to pacify Koguma by soliciting mounting: these are unmistakably subordinate behaviors. Jess, the son of the influential female Tare of the northern subgroup, at one period did the same thing repeatedly to the fourth-ranking male, Hachi.

This kind of behavior, from mid-adolescence into adulthood, may be the self-assertion, both physically and mentally, of a male approaching full adulthood. He is apparently starting to establish his position among the adult males. A strong mother, however, adds a complication. Her son achieves an unreasonably high position. This elevated rank may be a sham and a temporary phenomenon or it may be maintained. Interestingly, young sons seem to discern the capability and effect of their mothers very well, because the sons of low-profile mothers do not make these bold challenges.

Relationships Between Adult Males and Females

In general, males and females have two kinds of relationships. One is sexual, and the other is competitive, centering around food. Males and females have sexual interactions, and if these interactions result in conception and birth, their genes are passed on to the next generation. Consequently, sexual acts are reciprocal, cooperative work. Once partners are chosen and the pair enters into a sexual transaction, the act of mating is likely to be based on very intimate mutual feelings. Competition centering around food, however, is another issue. Because individual survival is at stake, intense hostility and aggression can develop between males and females.

In almost all higher primates, males have greater physical strength and locomotor capacities than females. The common chimpanzee is no exception, and when there is not enough of a desirable food, females may be at a disadvantage. Research by the Hasegawas, which covers a two-year period at Mahale, provides an example. Because there are shortages of the provisioned food (sugarcane), if a female happens to see some that has fallen under dense bushes and has gone unnoticed by males, she may pass it by, pretending there is nothing to eat in that place. But when the males move away, she will hurry back to the spot and pick up the sugarcane. Nevertheless, an ill-tempered male who has been keeping an eye on the female's behavior may fiercely attack her and steal the sugarcane she was holding.

At the Wamba feeding site, the picture for pygmy chimpanzees is completely different. Middle- to low-ranking males are at the head of the party when it first enters the feeding site. A little later, when the high-ranking males and females enter, the males pick up and carry at least one piece of sugarcane, and scatter into the marginal areas. They do not scatter owing to wariness of the researchers, but to avoid aggression from other males. Several high-ranking males linger at the feeding site, but the females do not seem to mind; they approach males, pick up sugarcane at their feet, and leave. Some females sit right next to the males and begin to eat. When they have plenty of time for sugarcane, males and females appear very much at ease with each other.

When a shortage of provisions is likely, even high-ranking males, on occasion, may be upset when a female approaches. They gather attractive morsels around their body and passively defend their sugarcane by turning their back on the female or by carrying away what they can and fleeing bipedally. Among middle- and low-ranking males, the problem is negligible; when females (especially middle-aged and old females that have stature) approach, the males readily get out of their way. Of course, there is some aggressive behavior by males toward females, and females have been so violently attacked that their ears were torn. These cases are rare, however, particularly relative to the frequency of male-male aggressive behavior (Table 23).

] 185 [

Anatomically, the pygmy chimpanzee is the least sexually dimorphic ape, but even so, the males are larger and stronger. As mentioned previously, the female's continuous sexual receptivity may suppress male aggression toward females. This explanation best fits the case of young females. When young females enter the feeding site, they look first for signs of males. Frequently, right after a male arrives at the feeding place, his penis becomes erect. When a female finds a male like that, she may approach him indirectly. The male looks at her, and if he gives a little sexual display, she will run to him and invite copulation. A young female may also approach directly and present, even though the male may not be looking at her. In either case, when copulation ends, she settles down and searches for food.

The sexual skin of female pygmy chimpanzees who wander from group to group is always maximally tumescent. Continuous sexual receptivity may be a necessary adaptation to protect females from males. Penile erection at a feeding site is a behavior often seen in captive chimpanzees when they are fed. E. J. Kollar, W. C. Beckwith, and R. B. Edgerton (1968) report that this phenomenon is a result of excitement and has no sexual meaning. But, if the aggressive impulse of the male toward the female is first suppressed by copulation under conditions where aggressive contact is apt to occur, then penile erection during excitement (the male's indication of being sexually receptive to the female) may be regarded as one behavior selected, under natural conditions, for the coexistence of males and females.

In addition to copulation, we saw male-female RR contact and mounting. These are infrequent relative to their occurrence in male-male encounters, but they do occur. They are not morphologically or behaviorally different from RR contact and mounting between males, but they do not accompany preliminary aggressive behavior and almost always occur in a peaceful context. These male-female RR rubbings or mountings occur most frequently in the same context as copulation, such as when the whole party becomes excited upon arrival at a good food source or at a party reunion. This pattern suggests that pseudocopulatory behaviors have a social function similar to that of copulation, and are apparently used as substitutes to decrease sexual tensions when females are not sexually receptive. Consequently, adolescent females, who are continuously receptive, exhibit RR rubbing or heterosexual mounting less frequently than do adult females (Table 24).

The frequency of grooming between males and females is extremely high. Generally, sexual partners do not engage much in grooming rela-

Table 24. Pseudocopulatory behavior in adults and adolescents
(based on 330 observation-hours at the feeding site, 1978–79)

Pseudocopulating pair	Mounting	RR contact	GG rub
Between males (adolescent-adult)	103	31	2[a]
Between a male and an adult female	39	12	0
Between a male and an adolescent female	2	0	0
Between females (adolescent-adult)	5	0	318
Total	149	43	320

[a]The males face each other while hanging on a branch; they sway their hips, which jut forward, from side to side while rubbing against each other with erect penises. This is a rarely seen behavior.

Grooming behavior at the feeding site.

tionships, however, and grooming does not occur immediately before and after copulation. In other words, grooming is a friendly behavior between males and females when they are not involved in a sexual relationship.

In short, pygmy chimpanzees have developed many behaviors that, under a variety of conditions, foster amicability in relationships between males and females. Among them, copulation probably has the greatest effect. Without this social effect of copulation, pygmy chimpanzee females could not have lengthened the estrous period at the expense of lowering their conception rate. Sterile copulation would be merely a waste of time and energy if not for its social merit.

In pygmy chimpanzee society, the female has about the same rank as the male. This does not result only from continuous sexual receptivity; middle-aged to old females experiencing detumescence of the sexual organs during the sexual cycle are more self-confident than the habitually swollen, young estrous females and the older females receive respect from the males. Two factors may influence the relationships between male and female pygmy chimpanzees. First, psychological factors, arising from the long period of dependency of the son on the mother, may be working on both males and females. Middle-aged and old females may simply see the majority of males as "childish," like their own sons. Second, there may be a "female bond." Several examples have been reported in which a male provoked a female and a group of females cooperated in a counterattack. A group of males will not attack a female, but the opposite can occur.

A female, with infant, grooming an adult male. This is the most common grooming association.

Familial grooming: a young adult male, Mon (right), is grooming his elder brother, Ibo (middle), who is grooming their mother, Kame (left), whose infant lies on her side.

When the victim is a low-ranking male, we expect that a high-ranking male is on the attacking side, and it becomes a serious matter. When the victim is a high-ranking male and other males are not involved, the victim can certainly be defeated by a mass attack of troublesome middle-aged females.

Relationships Between Adult Females

From the beginning, common chimpanzee research has revealed a low frequency of social interactions between females. Females do not associate with each other much and, in particular, they rarely engage in friendly behaviors such as grooming (Nishida, 1970). By contrast, female pygmy chimpanzees are very sociable. In particular, females with infants and juveniles (essentially all multiparous females) forage in close company, but aggressive interactions rarely occur. At resting time, grooming between females is common, although less frequent than between males and females (Table 25).

We cannot say, however, that aggressive interactions never occur be-

Table 25. Grooming interactions
(based on 330 observation-hours at the feeding site, 1978–79)

Grooming relationship	Number of cases	Proportion involving kin	Number of minutes	Proportion involving kin
Male-male	27(5)	0.19	426(112)	0.26
Male-female	103(31)	0.30	1,766(406)	0.23
Female-female	46(0)	0	393(0)	0
Male-juvenile/infant	56(31)	0.55	210(103)	0.49
Female-juvenile/infant	149(129)	0.87	614(534)	0.87
Juvenile/infant-Juvenile/infant	5(0)	0	4(0)	0

NOTE: Adolescent and adult male or female are given simply as "male" or "female." The numbers in parentheses show the number of cases and the amount of time that related individuals (mother-off-spring, siblings) groomed. Both the number of cases and amount of time that nonrelated individuals groomed are presented under "male-female," "female-female," and "male-male" relationships.

tween females. Occasionally, such as right after arrival at the feeding site, a female may jump on a neighboring female, drag and bite her, hold her down, and steal her sugarcane. Such attacks by females differ from those by males. Before physical contact, males glare at each other and usually progress through stages of threat in which they approach while glaring. Aggression between females, however, has the special feature of occurring suddenly, without any warning.

There are times when aggression between two females starts silently, but then screams rise as a fight develops and the two combatants begin rolling around on the ground. Other females may join, creating great confusion. In contrast, a fight between males is usually one-sided from start to finish, and is over quickly. Moreover, a female whose food was snatched and who was treated badly may roll on the ground on her back and scream in a "temper tantrum," whereas adult males do not express this kind of childish anger.

The pattern of antagonistic behavior between females strongly suggests that dominant-subordinate relationships (or the rank system) are underdeveloped in females compared to males. In general, relations between females are peaceful. Even when feeding close to each other, they are usually relaxed, and they rarely show expressions of submission such as screaming or grimacing.

The most conspicuous and striking social behavior between female peers is genito-genital (GG) rubbing, in which females rub against each other's external genitalia. The following is a typical example. First, female A approaches female B, who is in the midst of feeding. Female A, who approached in an unconcerned manner, lingers a bit and sits down at a distance, but within easy reach of B. Then, "a demand for invitation" follows, either by A or B. Let us suppose that A makes the invitation. A stands bipedally, extends her hand to B, puts her face close to B's face, and peers directly at her. If they are in a tree, A grasps a branch of the tree with one or both hands and peers at B while hanging above her. If B does not react, A may grasp B's knees with her feet, or some other body part, and shake her. Alternatively, A may touch B's shoulder and peer at B in a request. These are all requests that say, "Please associate with me in genital rubbing." Then, B rolls over on her back and spreads her thighs. When A starts to mount her, B embraces her ventrally as if she were a baby. A and B put their genitalia together and rub them against each other. If in a tree, one will hang from a limb above and open her thighs. Her partner will face toward the offered genitalia, and will either embrace or hang from the same branch as they begin to GG rub.

The part of the sexual organ that touches is the pointed part at the tip (clitoris). The quick, rhythmic thrusting differs from copulation, in that the hips sway from side to side. A female engaged in GG rubbing juts out her hips, and often has an expression (called "pout face") that conveys an uncontrollable emotion. She emits the same kind of scream as when copulating.

Two females genito-genital rubbing.

A female screams during GG rubbing.

GG rubbing is evidently comparable to copulation between female partners. The pygmy chimpanzee is the only one, out of 200 species of primates, to devise this behavior, and whenever they are delightedly absorbed in this "lesbian" behavior, they seem proud of their splendid invention.

GG rubbing often occurs in the same social context as copulation, rump-rubbing, and mounting, when excitement is elevated because of tension between individuals. The frequency of GG rubbing is highest for a while after arrival at the provisioning site, and we infer that this seemingly erotic behavior has the important social function of reducing tension between females. The effect of the GG rubbing may be that females are able to contact each other openly, even while crowded together. In this way, female pygmy chimpanzees raise the upper limit of their sociality and succeed in creating conditions for binding themselves into intimate social relationships within a unit group.

Aside from the unique nature of this behavioral form, the striking thing about GG rubbing is that it creates a harmonious stability in the relationship between two individuals, not by fixing a superior-inferior relationship as in a rank system, but by creating a condition of equality.

The females who approach and "beg" for an invitation to GG rub are younger and, consequently, many probably have lower status. Influenced by females who are senior in power, perhaps they obtain their own "sense of security" by relating through GG rubbing and soliciting "invitations."

Older, influential females do not perform dominance displays to show off their high position. (I call them "influential" because we do not use the word "high-rank" with respect to females.) These older females solicit various social contacts from their followers such as grooming, GG rubbing, and copulation. They seldom receive threats or attacks from others (including males), and also rarely display aggressive behavior. They are respected out of affection, not because their rank is high.

Curiously, when a female wants GG rubbing, an influential female will lie on her back, which tends to lead to mounting. When the females are close in age, they both present lying on their back, and may even do a compromise half-mount. If we were on the bottom being held down, we would probably feel submissive and inferior, but female pygmy chimpanzees seem not to take it that way. When GG rubbing, the female on the bottom takes the position of the female during ventral copulation. In this position, she looks proud and affectionate.

The only copulatory position in most primate species is the dorsal mount. In Japanese macaques and baboons, the male who takes that dorsal position is the dominant one, except in special cases. The female who is mounted is in a condition of nonresistance, facing away from her partner. The mounter has in his field of view the whole body of his partner, but the female, if she does not look back, cannot even see the facial expression of her mounter. By contrast, in the ventral copulatory position, there may not be an element of dominance-subordinance. Except for the physical differences of being on top or being on the bottom, the positions of both actors are identical and equal in status. Each can read the other's facial expressions.

The external sexual organs of a female engaged in GG rubbing do not always reach maximal tumescence. The larger her sexual organ and the more distended her swelling, however, the more attractive she is. An attractive female is often invited, and also often invites, GG rubbing. Consequently, the long estrous period of pygmy chimpanzees is not just for engaging in copulation, but also for GG rubbing. That is, it is necessary in creating stability in the relationships between females and in deepening friendships.

The external genitalia that are distinctive of female pygmy chimpan-

A female carrying a juvenile approaches a female with an infant.

The female who is standing solicits GG rubbing.

The two begin GG rubbing.

The two part company.

zees have a location and shape that are well-suited to females facing each other to rub their clitorides. I am convinced that this evolved for GG rubbing rather than for ventral copulation.

Immature and Adolescent Social Relationships

Because an offspring is carried by its mother almost all the time, during infancy there are few opportunities for others to contact the infant. About six months after birth, the infant extends its delicate hands to other individuals that approach the mother. These are the infant's first attempts to make active social contact. The majority of attempts of this kind are ignored, but occasionally, the infant receives a finger from somebody. When the mother is grooming someone else, the infant often crawls over the bodies of both adults, but they do not get annoyed or push it aside. The infant gradually increases its activity and broadens its behavioral repertoire. Elders are tolerant of behavior of any kind until the infant is about two years old.

Male infants are more precocious than females and begin to show sexual behavior in less than one year. When a mother finishes GG rubbing, her male infant clings to her partner and inserts his erect penis into her. Throughout the juvenile period, sexual behavior gradually becomes a regular activity. When the juvenile encounters his mother or other adults engaging in copulation and GG rubbing, he immediately runs and clings to either one's stomach or back, and screams. Then, when the adults conclude their activity, they embrace the juvenile and practice similar behaviors with him. There are juveniles that cannot wait to thrust; they cling to the male's hips during copulation and insert their penis in the female in the midst of GG rubbing.

Adult males in the midst of copulation sometimes get annoyed and push away clinging juveniles with their elbows, but adults are generally cooperative in awakening the erotic impulses of these juveniles. A female that has finished copulating and GG rubbing will stop and wait when a male juvenile clings to her behind; the female will grasp the juvenile's hips and insert his penis. Occasionally, a female will lower her hips to accommodate and make insertion easier. If so, the young male begins to thrust awkwardly but enthusiastically. At these times, other juveniles come and touch the female's behind, and wait for their turn. The female does her best to accommodate the partners, one by one, without much ado. Once in a while, the female will sit down and ignore them, but when this happens, the juveniles start to scream. The female may then lose patience and wearily lift up her hips as if to say, "he's just a hopeless child."

Adult males are also enthusiastic about "sex education." After copulation, they mount and thrust at juveniles, either male or female, that approach and present. Without inserting their penis, the adult males rub against the top of the hips and the thigh. Often, instead of thrusts, an adult male raises one ankle in the air and lightly scratches the side of the juvenile's chest.

Adult males are frequently the ones who invite the juveniles. A male may take a juvenile from its mother and thrust while it is clinging to his belly, or he may approach a solitary juvenile and mount. Carried away by enthusiasm, an adult may repeat this kind of behavior several times with the same individual. As a rule, one continuous bout lasts longer than true copulation, and examples of thrusting have lasted close to two minutes.

The frequency of sexual behavior in juvenile females is low relative to juvenile males (31 cases for females vs. 227 cases for males, Table 26), because unlike juvenile males, who vigorously engage in copulation with females, juvenile females hardly ever engage in GG rubbing with adult females. In contrast to males, who can have a penile erection from about six months after birth, juvenile females have extremely small external genitalia throughout the juvenile period. Only after they are about six years old and close to adolescence can females join in true GG rubbing.

Relative to adult sexual interactions, those of juvenile pygmy chimpanzees are rare (Table 26), which differs from other hominoid species. Juvenile pygmy chimpanzees do not need to get involved in immature sexual games with juvenile peers, because they have the good fortune of receiving direct coaching from experts on sex, the adults.

"Play" begins one year after birth, and with the exception of slow-moving play with the mother ("doll play"), it is solitary play. "Locomotor play," in which youngsters hang and clamber in the branches, is representative. After one year of age, they begin to play with other infants, but basically as an extension of locomotor play in which peers are together.

During the juvenile period, individuals enter into ordinary "social play," consisting of chasing each other from branch to branch and grappling and wrestling. They do this tirelessly, generally during the adult rest period. In one game, they jump into the nest, wrestle for a while, and leave. Then, after running around, they jump into the nest again. Long-term play occurs in the trees, but play on the ground rarely lasts long.

Table 26. Pseudocopulatory and copulatory behavior
among immature individuals (juveniles, infants)
(based on 330 observation-hours at the feeding site, 1978–79)

Position and kind of sexual behavior	Partner						
	Adult male	Adoles-cent male	Adult female	Adoles-cent female	Imma-ture male	Imma-ture female	Total
Immature males							
Was mounted dorso-ventrally	32	10	1	0	7	0	50
Mounted dorso-ventrally	7[a]	1	39	27	—	1	75
Ventro-ventral copulation	11	7	39	20	11	2	90
Copulation with position change	0	0	1	1	0	0	2
Copulation in undetermined position	2	0	5	0	0	0	7
Genito-genital rub	0	1[b]	3	0	0	0	4
Rump-rump contact	1	0	0	0	0	0	1
Immature females							
Was mounted dorso-ventrally	7	5	0	0	—	0	12
Ventro-ventral copulation	6	2	0	0	—	0	8
Copulation in undetermined position	1	0	0	0	—	0	1
Genito-genital rub	0	0	3	4	—	0	7
Undetermined sex							
Ventro-ventral copulation	1	0	0	0	0	0	1
Total	68	26	91	52	18	3	258

[a] In three cases, an adult male was mounted and thrusted against when he was in the midst of dorso-ventral copulation (multiple-individual copulation).
[b] A behavior of rubbing with the penis (see Table 24).

Juveniles often use small branches in play. One individual will pick up a small branch and run away, signaling "try and get it." With a playmate in pursuit, he dashes around, passes the branch in a baffling way from hand to foot to mouth, and puts it between his thigh and abdomen. Finally, when the branch is snatched from him by force, he changes roles to chase his playmate. I think that the origins of tag and all ball games are very old. Although adolescent females join in these chase games, multiparous females do not engage in play, except for playing with their own infant. Small branches are not the only "tools" of play; sugarcane also is often used.

When adolescent and adult males join in play, the game is almost always play-wrestling. Play-wrestling occurs not only between juveniles, but also between adolescent males and between adolescent and adult males. As expected, play between adult males is rare. When a senior and a junior are engaged in play-wrestling, the junior is invariably on the aggressive side. His elder squats on a branch or rolls over on his back; then, junior attacks a favorite spot by "play-biting," grabbing and biting a hand or a foot. Whenever we hear the "play-pant," a quick "hat-hat-hat," we know that somewhere someone is play-wrestling.

Juveniles often engage in "food-begging" from individuals other than their mother. The success rate, however, seems to be poor, and the juveniles often give up after a short time.

An adult male may embrace an unrelated infant or carry it on his back. To do this, he approaches a female carrying an infant and stares at the infant while extending his hand. The infant may jump on his back, and if the infant clings to his stomach, he will quickly leave the mother. The mother will tolerate a separation of about 5 m, but if they go farther away or if the infant begins to whimper, the mother will hurry to take it back. No cases of females engaging in this kind of motherly behavior have been reported.

The behavior of elders toward juveniles (2 to 4 years old) may be summed up in a single word, tolerance. There are no reports of severe scolding or violent attacks directed at juveniles. However, aggressive behavior toward juveniles does occur often as they approach adolescence. Elders begin to chase the older juveniles, to take their sugarcane, and to threaten them. In serious cases, they even hold them down and bite them. Because their mothers rarely come to their aid, males and females approaching adolescence begin to spend more time alone on the periphery.

Adolescence (7 to 14 years old) may correspond to puberty in humans, which extends from 10 to 15 or 16 years of age. It is the transitional period from childhood to adulthood. Although there are exceptions, usually a male at this time lives at the periphery of the group. He moves with his mother, but often he separates from all other members of the group. He keeps to himself, especially during times of excitement and enthusiastic feeding, because when he is in the center of the group, he will often be threatened or attacked by other males. No longer a juvenile, he is not tolerated.

Adolescent males keep company with their mothers during the rest period. They finish their meals and lie sprawled, separately scattered about. They make their nests, and when they rest in the shade, the adolescent males casually approach their mothers and often enter into grooming relationships. Almost all of them are inferior to any adult male; they rarely mount and rump-rub with adult males and are still not recognized as "fully adult."

Among adolescents, dominant-subordinate rank is apparently not yet established. Even in postadolescent males, attacker-victim relationships are not fixed. Also, aggressive interactions among adolescents are not violent and frequently transform to play. An adolescent male may branch-drag, but he does this infrequently compared to an adult male.

Already mentioned in passing, "peering" behavior, a common behavior in adolescents regardless of sex, will be formally introduced here. When peering, a pygmy chimpanzee sits or stands quadrupedally and gazes intently, from the front or side, into another's face. The actor gets close enough to almost make face-to-face contact and, with a calm facial expression, stares at the other. This stare usually lasts one to two minutes, but one case lasted five minutes.

Peering is typically a behavior that juniors direct toward their age-mates or seniors. The one being stared at usually pretends not to notice, but some kind of social interaction occurs when they are in each other's line of vision. The effect differs according to age, sex, and rank, but the typical behaviors that are generated include copulation, GG rubbing, play, and food sharing. Peering rarely leads to aggressive interaction, from which we deduce that it is not threatening.

A mother allows her infants and juveniles to come close to her face. While they stare, "food-begging" often occurs, and peering may be an extension of this behavior. Peering, however, is not just food-begging, but instead it seems to be begging for any kind of friendly social contact. The frequency of peering behavior increases with age, adolescence being the peak period. It is hardly ever seen in males after middle-age, but even after reaching old age, females use peering to solicit GG rubbing.

Adolescent females are unique beings within the group. As immigrants from other troops, they are newcomers from the point of view of other members. Their behavior is always reserved; they keep their distance from all disputes; and they seldom emit loud vocalizations, even when exchanging calls with another party. They rarely resort to aggressive behavior of their own accord, and if attacked, they just scream and flee without counterattacking.

A juvenile peers at an adult male.

Peering behavior.

An adolescent male peers at an adult female.

These young nulliparous females are not completely socially inactive, however. They respond to the behavior of males. Because they quickly approach and present when they perceive even a slight sign of courtship display, copulatory frequency is highest in these adolescent females. Also, they eagerly approach older multiparous females to "peer" and to solicit GG rubbing. After engaging in these kinds of behaviors for a while, the nulliparous females carry their food and retreat to a safe feeding spot in a tree on the outskirts of the feeding site. During the rest period, however, they re-appear, approach the older females, and begin to groom. These adolescent females spend much more time grooming than being groomed. During the rest period, they "play-chase" with the juveniles for a long time.

Thus, the young nulliparous adolescent females avoid aggressive contact on the one hand and initiate friendly contact on the other. They seem to be working to stabilize their own position in the group by establishing friendly social relationships, rather than by building dominant-subordinate relationships. They seem to know well that they have no supporters or guardians in the group. This reminds me of a new Japanese bride who bravely enters the groom's large household. If she marries the oldest son, she may also have to live with her husband's parents and the unmarried brothers and sisters of her husband. Because she is picked on by in-laws, and her own family is distant, the new bride is often under great stress. She must be very patient.

Female adolescence may be roughly divided into three stages. In the first stage (7–8 years old), the female has just left her natal group. Her sexual organs are small, unlikely to be penetrated by an adult male penis, and her body also is small. Nevertheless, she actively engages in sexual and social behavior. Sometimes, however, a little childishness remains, and she may become absorbed in playing by herself for a long time. When she enters the middle stage (9–12 years old), the sexual organs increase in size by swelling. During this period, copulation and GG rubbing occur most frequently. The last stage (13–14 years old) lasts about nine months, from conception to birth of the first offspring. Copulation occurs up until about one month before delivery. The sexual swelling gradually slackens up to the time of delivery, at which time the female becomes socially and sexually inactive.

We conclude from the data obtained so far that the inclination toward a specific unit group deepens during the first half of adolescence. By the time a female is 12 or 13 years old, her position within the group seems to be almost settled.

Relationships Between Groups

More than 60% of the home range of Wamba's E-group overlaps spatially with the home ranges of other groups. In terms of time, the percentage of overlap is smaller, but, at least spatially, the home range can be shared with other groups. Rarely do groups have direct contact, and we have obtained few data about mutual interactions between groups.

K-group and B-group have come close to each other. K-group often forms large parties, with as many as 70 chimpanzees. At those times, without exception, they were eating *Dialium* fruit while making a great commotion. Once I saw B-group cross the road just 200 m away. About 30 members in B-group had been traveling silently on the ground, but when they were 500 m from K-group, they climbed a short distance up the trees and began a loud chorus. By so doing, the smaller group seems to avoid contact and probably keeps the frequency of group contact low.

In some cases, the larger group may pursue the smaller group. One evening, for example, a small party from E-group went south, but on their way to the feeding site, they saw P-group and started to retreat. Even though it was getting dark, P-group followed E-group that day for more than a kilometer. E-group does not always avoid P-group, however. When the northern and southern subgroups of E-group join together and form a large party, they pursue P-group. Occasionally, group encounters develop into conflicts. Kitamura reported that when E-group and P-group met at the feeding site, a violent fight occurred, leaving several individuals injured.

In contrast, a quite different scene was observed when K-group and E-group made contact. When artificial food (sugarcane) was provided, the peripheral members from the two groups intermingled and fed on it. At that time, a middle-aged male of E-group was once observed chasing a young adult male of K-group. It would be an exaggeration to call this an intergroup conflict, however, because neither male was supported by any group member close to him. This case was more similar to an intragroup aggressive encounter in which a higher-ranking male scolded and drove away a young male who had decided to enter the feeding site. According to my research associate B. Mulavwa, one day several individuals, including young adult males of E-group, stayed behind in K-group for as much as 20 minutes, but nothing happened.

Another intergroup encounter between E-group and P-group is described below. When a small party of E-group was at the artificial feeding site, another party nearby erupted in chorus. Immediately, the members of E-group responded with loud vocalizations and left the feeding site, heading toward the sounds. Because this kind of response was common in normal party reunions, I remained at the feeding place expecting that soon the members of E-group would return. This time, however, the clamoring voices continued for two hours, with no sign that they would return. I approached the parties and found E-group and P-group close to each other, incessantly exchanging calls, but without apparent aggression.

Unit group contacts are clearly diverse. Because grooming, copulation, or GG rubbing between members of different groups have not been observed, relationships between groups cannot be generally described as friendly. It does not follow, however, that aggression, flight, or fighting will develop, nor can we easily attribute these interactions to antagonism. Moreover, because young females emigrate when different unit groups draw near or make contact, the overlap of group home ranges increases the opportunity of young females to emigrate. Consequently, overlap is a phenomenon favorable to maintaining social structure.

Comparison with Common Chimpanzees

When we compare social relationships of the pygmy chimpanzee and the common chimpanzee, we recognize differences between and within all sex and age classes. In both species, kin are bound together by friendly mutual interactions such as food sharing, grooming, cooperative play, and care behavior. This is especially true between mother-offspring dyads. The differences between the species appear when the sons and daughters approach adolescence. Not long after entering adolescence, female pygmy chimpanzees separate decisively from their mothers, never to return. In common chimpanzees, the daughters also leave, but at Gombe Park, a lineage of females spanning three generations lived within the same group; separation apparently was not so stringent as in pygmy chimpanzees. Also, at Mahale, a female who had emigrated from a group returned once to that group.

In adolescence and adulthood, male pygmy chimpanzees accompany their mothers. In contrast, adult male common chimpanzees are incorporated into a "male band" and quickly become estranged from their mothers. According to Jane Goodall (1971), a mother occasionally will display fearful behavior toward her own adult son. This has not been observed in pygmy chimpanzees.

One difference during infancy is recognized. Among common chimpanzees, mother-son mating is commonly seen. If immature mating is necessary for the development of normal mating behavior, then where the mother-infant unit often lives independently of other members, as in common chimpanzees, practice mating with the mother may be necessary. On the other hand, among pygmy chimpanzees, where there is no lack of unrelated sexual coaches, mother-son mating is infrequent. In pygmy chimpanzees, the early establishment of incest avoidance between mother and son may be related to the sons' "mother-complex" (close mother-son relationship).

In both species, the rank system founded on dominant-subordinate relationships may control the social relationships between unrelated males. Between male common chimpanzees, however, the rank relationships are clearer, and friendly relationships, such as grooming, are stronger. Males form a solid mutual bond, and they forage together. In contrast, between male pygmy chimpanzees, clear rank relations do not exist, and friendships are weak. The long-term dependence of male pygmy chimpanzees on their mothers probably accounts for this difference.

Among common chimpanzees, all adult males dominate all adult females. By contrast, dominant-subordinate relationships of male and female pygmy chimpanzees vary depending on the individual and the conditions. This is probably the reason that male rank within the group is unclear.

In common chimpanzees, the period of association between males and females is relatively short. Estrous females, who follow male groups, conceive after several sexual cycles; then after giving birth, a female accompanies her infant for a long time and lives independently of males. A female pygmy chimpanzee, on the other hand, spends time with the male throughout the sexual cycle as well as during child-rearing.

Female relationships are the most disparate between the two species. Among female common chimpanzees, mutual social interactions, such as grooming, were found to be rare since the beginning of research. In contrast to the asocial female common chimpanzees, female pygmy chimpanzees are extremely sociable and always feed, travel, and rest in clusters; social interaction between them is also lively.

In summary, unity among male common chimpanzees is stronger than that among male pygmy chimpanzees. In addition, relationships between males and females and among female pygmy chimpanzees are much friendlier than those of common chimpanzees. These differences are reflected in social behavioral differences between the two species.

There are definite differences between the two species in their agonistic behavior. In common chimpanzees, a behavior called "pant-grunt" signifies low rank and acts as an appeasement. The behavior consists of stepping forward and falling prostrate in front of a high-ranking individual, grimacing, and panting intensely, and it frequently occurs between adults, both males and females. In juveniles, a behavior called "bobbing," in which the body is prostrate and the individual moves its head violently up and down, may be a forerunner of the pant-grunt. A high-ranking individual reacts to pant-grunt with a pacifying behavior, such as an extended hand or an embrace that eases the tension of the low-ranking individual. Various types of pacification behavior have been reported at Gombe Park and Mahale (Goodall, 1968). Thus, common chimpanzees often go through a sequence of behaviors (dominance → appeasement → pacification) showing that dominant-subordinate relationships are an important element in the maintenance of common chimpanzee society.

In contrast, behavior similar to the pant-grunt is not seen in pygmy chimpanzees. When a pygmy chimpanzee is attacked, it grovels and shrieks violently, but these are merely manifestations of distress, not appeasement. Moreover, pacification behavior directed at a low-ranking individual by a high-ranking pygmy chimpanzee does not take the same form as in common chimpanzees. After the low-ranking member is attacked, it rarely waves, extends its hand toward the attacker, or makes a cry-face; likewise, the attacker has not been observed to pacify by shaking hands or embracing.

Thus, in pygmy chimpanzees, many behaviors thought to derive from sexual behavior regulate the relationships between non-kin. These behaviors maintaining pygmy chimpanzee social relationships vary depending on the partner's sex-class association and differ greatly from the pant-grunt and pacifying behaviors of common chimpanzees, which are used irrespective of the actor's sex. In the pygmy chimpanzee, rump-rubbing and mounting are behaviors used to maintain relationships between males and between males and anestrous females; GG rubbing is used only by females (Table 24). Copulation regulates the relationships between males and estrous females. In general, behavior that employs the special characteristics of the respective sex is used to regulate interindividual relations. This kind of "behavioral specialization" probably guarantees a greater effect.

A second point of difference between the two species is that the behaviors regulating the individual relationships of common chimpanzees are based on hierarchical relationships, whereas those of pygmy chimpanzees are based on mutual equality. If we exclude mounting, there is no difference between partners in form. Even in mounting, the subordinate is on top in close to one-third of the encounters; dominant-subordinate relationships do not always determine position. The coexistence of unrelated pygmy chimpanzees does not depend on rank or dominance and subordination, but instead it is based on and maintained by equality in the relationships.

The differences between the two species in relationships between adults and immature individuals are thought to be few. Jane Goodall (1971) stated that small infants of the common chimpanzee were shown almost unlimited tolerance from all other members of the community (unit group). At that time, however, "infanticide," the fatal aggressive behavior shown by adults toward infants, was not yet known. Infanticide was first observed by A. Suzuki (1971) in the Budongo Forest. An adult male in a tree was eating an infant that was still alive and crying weakly. Suzuki's report concluded that this was very abnormal, an example of deviance, and he showed that the common chimpanzee also displays madness. Subsequently at Mahale and Gombe, however, researchers began to observe a sequence of cases of infanticide, and they decided that the examples were not abnormal. Males and females participated in infanticide, but males were apt to kill males and females were apt to kill females (Goodall, 1977, 1983). Also, more male infants were killed than female infants (Kawanaka, 1981). The mothers of killed infants were newcomers to the group (Goodall, 1983).

Sociobiologists offer the following explanation for infanticide. The father of an infant who accompanies a newcomer is probably a male from another group. From a male's perspective, the infant, if female, may one day become an associate who might transmit the male's genes to the next generation; if male, the infant is merely a potential competitor and, consequently, he is decisively removed. Because suspension of lactation upon death of the infant allows the bereaved mother rapidly to resume menstruating (and to become capable of conception), the male benefits from infanticide by increasing the number of females available as mates; by consuming the infant, he also gains nutrition to support his own individual existence. Infanticide by females can be explained as interference in the reproduction of a competitor. The invested energy lost by the mother during the period from copulation, through pregnancy and the birth of the child, until it is lost is immeasurable. As such, infanticide is a threat to the female's reproductive success. It is thought to be a loss in terms of the continuation of the species, but since natural selection is a process that acts only on the individual, it is not necessarily a loss to the species.

The threat of infanticide may explain the greeting behavior that a female common chimpanzee who has just given birth demonstrates toward males. She repeats, with unusual enthusiasm, the "pant-grunt" to each male, one by one. In Toshikazu Hasegawa's words, it seems as though she wants the male who is the father to recognize his child. The apparent ineffectiveness of this desperate endeavor can be inferred from the many infants who were unaccounted for two to three years after birth.

On one occasion, Mariko Hiraiwa-Hasegawa observed several males surround a female who crawled on the ground and concealed her infant, while she pant-grunted fervently. Nevertheless, the villainous males attacked her one by one and seized the infant. On seeing this, Hasegawa momentarily forgot her position as a researcher and, brandishing a piece of wood, she intervened and confronted the males to rescue the mother and infant. The independent existence of a female with offspring, typical of the common chimpanzee, may be a defense of her children against males.

Whether or not infanticide occurs among pygmy chimpanzees is still uncertain. At Gombe Park, infanticide by common chimpanzees was first confirmed ten years after the beginning of observations. At Wamba, however, except for an infant that apparently died from malnutrition, all identified infants are being well-raised, and there have been no reports of missing individuals. Furthermore, these pygmy chimpanzee infants do not live in isolation with their mother; they are raised within a group. Female pygmy chimpanzees who have just given birth are mostly inactive and do not display an especially wide variety of attitudes toward males and other members. On the basis of these observations, the possibility of habitual infanticide in pygmy chimpanzee society seems remote.

In the common chimpanzee, even multiparous females transfer between groups, and their position is unstable. In the pygmy chimpanzee, however, adolescent females hold stable positions in fixed unit groups. When an adolescent female is pregnant, even if a male in her unit group does not have confidence that he is the father, he does sense that the father belongs to his group. That is, he has some confidence that a male of another group is not the father. This may be one of the factors that checks infanticide in pygmy chimpanzee society.

Male common chimpanzees display violent agonistic behavior toward members of other unit groups. At Gombe Park, a group that was provisioned from 1970 to 1972 split in two, forming the Kasekela group and the smaller Kahama group. Over the next five years, six of the seven Kahama group males were attacked and killed, one by one, by Kasekela group males, rendering the Kahama group practically extinct (Goodall et al., 1979; Goodall, 1983). In another example, the number of males in Mahale's K-group has decreased, and their near extinction is presumed to have been caused by attacks from the powerful neighboring M-group (Nishida et al., 1985). The male bond that tightly unites the group is a needed coalition for territorial defense. These males patrol the boundaries of the group territory, and occasionally invade deep into the territory of a neighboring group. Their behavior has been called "patrol" or "scouting" behavior.

In the unit groups of common chimpanzees, the number of adult males is considerably less than the number of adult females because of the loss of males, first, from infanticide and, then, from intergroup conflict. By contrast, in the habituated group of pygmy chimpanzees at Wamba, no infanticide has been observed and no adult males were killed from 1976 to the beginning of 1985, except for one that was shot by a hunter. The adult sex ratio of the group remained at about 1:1. This implies that the relationships between groups of pygmy chimpanzees are more peaceful, or, at least, that pygmy chimpanzees do not go so far as to kill each other. Judging from the lack of evidence of intraspecific killing throughout our long-term observations, we believe that the pygmy chimpanzee lives in a much more peaceful and mutually tolerant society than the common chimpanzee, its sibling species.

CHAPTER EIGHT

Why Are Pygmy Chimpanzees Interesting?

According to mitochondrial DNA analysis, the time of divergence of the pygmy chimpanzee and the common chimpanzee is estimated to be 1.5 mya. From a geological perspective, this is extremely recent. Clearly, the two types of chimpanzee have many basic features in common, such as morphology, ecology, society, and behavior. When we delve into deeper levels of comparison, however, we find unexpected differences between the two species, which diverged practically yesterday.

First, the pygmy chimpanzee is relatively stenoecious (has a narrow ecological niche) while the common chimpanzee is more euryecious (tolerates a wide range of habitats) (see Chap. 2, Distribution). As judged by the size of its range of distribution, the common chimpanzee is by far the more successful species. That differential success comes from the differences in the habitats occupied by these two types ever since they were separated by the Zaire (Congo) River. The ancestors of the common chimpanzee were on the north side of the Zaire River and were apt to experience the effects of climatic change and to be continually tested by severe environmental changes. The common chimpanzee's relatively broad adaptive capacity is probably a result of this. On the other hand, the ancestors of the pygmy chimpanzee were forced into a smaller region, south of the Zaire River, and had their roots in a comparatively stable environment.

That pygmy chimpanzees are relatively stenoecious while common chimpanzees are more euryecious is consistent with the ecological data. Common chimpanzees have developed such specialized behaviors as tool use, and have acquired the greater ability to exploit their resources, which is also an adaptation to a harsh environment. By contrast, pygmy chimpanzees live in relative affluence, spending the majority of the year feeding on the great quantity of fruit in the midst of the forest. If pygmy chimpanzees arose in a relatively rich environment, it was unnecessary for them to develop skills in resource exploitation comparable to those of common chimpanzees.

In the domain of social skills, however, pygmy chimpanzees excel at individual compatibility and sociability. They have developed various social behaviors in relation to these skills, and may have even modified their sexual behavior. Surplus energy that under harsher conditions would have been spent on merely surviving—on the everyday activities of searching for and consuming food—has been directed into affiliative behaviors.

The fission-fusion nature that characterizes societies of common and pygmy chimpanzees is probably a social adaptation to the erratic ripening of the fruits that constitute their principal food. In the tropical rain forest, fruit production is unstable, with the quality and quantity of the food supply changing both spatially and seasonally. The size of temporary foraging groups (parties) varies according to the amount and the pattern of distribution of foods so as to maximize feeding efficiency. In general, chimpanzees search widely for food and are able to use efficiently food resources of many kinds.

The relationship between skills in resource exploitation and the make-up of the environment may be understood more clearly by examining another African ape, the gorilla, which mainly feeds on the stems, shoots, pith, and leaves of wild herbs. These food sources are abundantly and evenly distributed throughout the gorilla's home range and fluctuate little throughout the year (except for a few special items such as bamboo shoots). With this pattern of food supply, the aggregation and dispersal patterns of individuals will not increase feeding efficiency. What is desirable is to divide the habitat into small parcels frequented by unit groups of small stable size. The unit group of the gorilla takes just that form.

That pygmy chimpanzees form larger size parties than the common chimpanzees of Gombe and Mahale can probably be explained by differences between the patterns of food supply of the two species. The principal food of pygmy chimpanzees is located in the crown of tall forest trees and is produced abundantly. Probably often the amount of food supplied at one individual food tree or at one locality is greater for pygmy chimpanzees than for common chimpanzees. Because food trees of pygmy chimpanzees are distributed non-uniformly throughout the forest, the source of food changes spatially; it also changes seasonally. Thus, faced with scattered but concentrated food sources, pygmy chimpanzees improve their efficiency more by forming large parties than by separating into small parties. Moreover, in the presence of a concentrated supply, several unit groups may hold the same food source in common, which again improves the efficiency of food collection. The presence of a concentrated supply may explain, in part, the large overlap between the home ranges of neighboring groups of pygmy chimpanzees.

Because the number of individuals seems to be one of the important factors determining the relative status of groups, large party size is advantageous. To maintain a large size party, however, individuals must be able to coexist. Generally speaking, sexual behavior (while one is actually engaged in it) has the power to build unconstrained male-female intimacy. Although nonreproductive mating is common in primate societies that form polygamous groups (Small, 1988), in pygmy chimpanzees sexual behavior is also used to establish intimate relations between individuals. Relations are formed not only between a male and a female but also between members of the same sex. Indeed, the evolutionary path taken by pygmy chimpanzees, one that encourages coexistence, has proven a success.

While engaged in sexual behavior, one is temporarily defenseless to an enemy. Because of this drawback, the manifestations of sexual behavior are curbed to what is minimally necessary in many animals. In pygmy chimpanzees, however, the apparent use of sexual behavior in a purely social context led to a rise in the frequency of occurrences of sexual behavior. As the demand for sexual activity increased, the period of sexual receptivity of females lengthened, accomplished by the prolongation of estrus in cycling females and the shortening of anestrus after parturition. In modern pygmy chimpanzees, the sexual receptivity of females is semi-continuous.

Frequent copulations resulting from the semi-continuous sexual receptivity of females might lead to the shortening of the birth interval. The subsequent increase in the number of dependent offspring would be a great burden on females. Among pygmy chimpanzee females, this burden is effectively avoided by the dissociation of estrus and the ability to conceive. That is, after parturition, the resumption of the swelling cycle appears to come earlier than the resumption of the menstrual cycle. This is inferred from the observation that many females resume copulating with males within one to two years after parturition but do not give birth until an average of five years after parturition. Thus, many of their copulations do not lead to conception.

In contrast to the pygmy chimpanzee, the common chimpanzee, which inhabits a harsh food environment, naturally came to form a different kind of society. For example, in Tanzania, the forest occupied by common chimpanzees is not prodigious. There are fewer large trees and the quantity of fruit produced by one tree is much smaller than in the Zaire Forest. In that kind of habitat, common chimpanzees cannot survive by using only the "concentrated" food supplies that suffice for pygmy chimpanzees. Instead, they must search diligently for their food, including the small food sources that are scattered evenly throughout their home range. No single social structure is able to efficiently use both kinds of food sources at the same time, because the patterns of supply are diametrically opposite. In order to locate and exploit concentrated sources of foods, a large number of individuals need to forage together over a wide region, whereas small quantities of scattered food stuffs can be obtained more efficiently by small numbers of individuals searching thoroughly over a narrow region.

Common chimpanzees, pressed by necessity to use food sources of both kinds (concentrated and scattered), developed a peculiar dual social structure. First, the males that had excellent locomotor faculties and the agile estrous females without infants gathered together, and took charge of the large concentrations of food. From this "sexual" grouping, the babies were born and raised. Second, the mothers, slowed by their infants, split up from each other and entered into independent small home ranges. These mothers exploited food sources in small supply and paid close attention to child-rearing.

In order to maintain this dual social structure, the social nature of females, and especially sociability among females, must be restrained. The same behavior that would bring about intimacy in females would later become an obstacle to their living separately. That obstacle has been bypassed by lowering to a minimum the investment of energy into sexual behavior. Copulation ends after several seconds, and estrus is directly linked to the ability to conceive. A few menstrual periods after the resumption of the estrous cycle, the female is pregnant. By the time of delivery, the sexual organs completely deflate, and copulation becomes impossible. Postpartum anestrus continues for three to four years while the female rears her offspring.

Among common chimpanzees, young females who are in estrus but experiencing adolescent sterility are not as attractive to adult males as are fertile adult females in estrus. This is further evidence that mating among common chimpanzees focuses on conception. This mating system, however, by delaying the establishment of intimate relationships between males and females, produces a problem for young nulliparous females, who will have difficulty attaching to a fixed unit group. As a result, a male does not know whether or not a particular infant is his. The lack of confidence in paternity and the estrangement of males and females arising from the generally slow development of sexual behavior may cause infanticide by males. If, however, the threat of infanticide forces the female with dependent offspring into a solitary life in her small home range, then for the common chimpanzee, even infanticide may be considered a necessary adaptive phenomenon that maintains their species-specific social structure.

The complete tyranny of males over females may have originally forced the female and her offspring to become "solitary," but subsequently, they may have preferred to live in quiet harmony by themselves. Thus, the common chimpanzee's society and its sexual relationships encourage, or at least do not prevent, the independent life-style of the mother and offspring.

Although dividing the home range into small mother-offspring units does raise food-intake efficiency in a harsh environment, it has also produced another social change. In this system, a large territory encompasses many mother-offspring unit boundaries. For males, a large territory increases not only their food base but also the number of available females (Wrangham, 1979). Therefore, the expansion of the males' territory and the maintenance of it become a top priority in the common chimpanzee. The relationship between neighboring groups becomes antagonistic; the solidarity of males within the same group becomes firm, and they form coalitions having a common goal.

By contrast, female pygmy chimpanzees and their offspring associate and move together more often. Pygmy chimpanzees seem to have a female-centric society, in which the males do not dominate or lead the group; they just follow. In this kind of society, the enlargement of territory is not at all tied to the acquisition of females. In pygmy chimpanzee society, the relationship between neighboring groups is not as hostile as in common chimpanzees, although there may be some antagonism. Infanticide and mortality during intergroup fights between males are also aspects of and agents of natural selection in common chimpanzee society. Neither seems to be a regular part of pygmy chimpanzee society.

Because common chimpanzee females do not belong to fixed unit groups and because they leave from and return to their natal group, they may mate with relatives. Thus, the inbreeding coefficient of the group is higher than that of pygmy chimpanzees. This implies that the frequency of homozygosity within common chimpanzee unit groups will be higher, and the expression of recessive traits will be more frequent. As a consequence, selection pressure is probably also stronger. If there are greater selection pressures on common chimpanzees than on pygmy chimpanzees, operating perhaps against offspring (infanticide), males (intergroup killing), and a group of recessive genes, then the speed of natural selection, and consequently the rate of evolution, may be greater in common chimpanzees.

The question is often raised, "which of the two types of chimpanzee is primitive and which one is specialized?" Some of the characteristics of pygmy chimpanzees are probably more primitive. Certainly, pygmy chimpanzees retain a larger number of ancestral genes because of their slower rate of selection. Therefore, perhaps we should look more to pygmy chimpanzees than to common chimpanzees for the various physical and behavioral characteristics of the forest-dwelling common ancestor of chimpanzees and man.

From this common ancestor, the ancestors of man diverged first, moving to the savanna where, under severe environmental pressure, they began to change at a tremendous evolutionary rate. Next, the ancestors of common chimpanzees branched off the forest-dwelling line and came, independently, to be exposed to the severe savanna environment. Because the divergence time was much later, however, the differences between them and the forest-dwelling pygmy chimpanzees probably still were not great.

The largest behavioral difference between humans and chimpanzees is that, in humans, the males participate economically in the mother-offspring unit in the form of a pair-society (the so-called human family), which serves as the basic social unit. This is in contrast to both species of chimpanzee, which have multi-male and multi-female groups as the basic social units. The common ancestor of chimpanzees and man, however, must have had a society that included the potential for either of these forms.

If we assume that the rate of change in the human line was exceptionally fast, then the society of our common ancestors may have been more similar to the chimpanzee form of society. There are several possible solutions to the problem of deriving a human family from a chimpanzee form of society.

First, the multi-male, multi-female groups must break down. For permanent male-female pairs to be maintained in a society, the interference of promiscuity must be overcome. Nevertheless, chimpanzees and many humans even now do not pass up a chance to be promiscuous. Because the subjects of immorality and fidelity are ancient, they are reflected with deep feeling in legends and tales, in religion, and in social norms. The importance given to these subjects suggests that humans, even now, are fundamentally promiscuous.

It is difficult to imagine that male-female pairs were differentiated within a group of promiscuous individuals while the integrity of the whole group was maintained. Multi-male, multi-female groups normally dissolve, however, in response to environmental pressures. It is easy to imagine that the origin of the family may have resulted from being driven into a corner under conditions in which multi-male, multi-female groups were subdivided spatially into pair forms.

The ancestors of man lived on the savanna, and there may be a society of common chimpanzees now occupying a similar place and facing environmental pressures similar to those of our ancestors. Common chimpanzees have a dual-form society in which males move about in a wide territory and exploit concentrated food supplies, whereas each mother-offspring unit lives in a narrowly partitioned independent home range and exploits a scattered food base. If the severity of the environment were to increase again, then the concentrated food supply could hardly be used at all, and the males would segregate themselves, one by one, into small home ranges. At this time, individual male and individual female-offspring units would bind together to ensure breeding.

The possibility that the human family originated when males joined mother-offspring units cannot be dismissed. There is, however, one extremely awkward element in this scenario. For males and females to form permanent pair relations, presumably through the medium of sex, females must always be sexually receptive. To the contrary, female common chimpanzees are anestrous for a long time until their offspring can take care of themselves. In common chimpanzees, females cannot use sexual activity to establish pair bonds with males at the very time they most need male assistance, during child-rearing.

It seems more natural that males and females had lived together from the beginning, as pygmy chimpanzees do, than that males joined females and offspring who had lived independently. Moreover, the continuous sexual receptivity of females could not have developed after moving into the severe savanna environment. Because copulatory behavior has the disadvantage (from the viewpoint of individual survival) of making one vulnerable to enemies, the notion that continuous sexual receptivity arose in a severe environment, such as that of common chimpanzees, ought to be dropped.

The continuous sexual receptivity of human females is presently believed to have been achieved long before human ancestors advanced into the savanna. As seen in pygmy chimpanzees, extended periods of female sexual receptivity can develop where food is abundant and predators are few, as in the forest. Early hominid females may have retained the characteristic of continuous sexual receptivity as they advanced into the savanna. There, females may have used their continuous sexual receptivity to draw males into the female-offspring unit. According to this model, common chimpanzees, as they adapted to the savanna, came to be "dehumanized" by sacrificing female sexual receptivity.

Now let us look back on the mother-centric family unit of pygmy chimpanzees and assume that permanent sexual receptivity of females has been achieved. Faced by a very difficult environment, the mother-offspring units of present-day pygmy chimpanzees would break up, thereby initiating an increase in feeding efficiency. When daughters attain full maturity, they leave their natal group, but sons, even mature ones, remain in the mother's unit. A young female leaves her natal mother-son unit during adolescence and is introduced into another mother-son unit. The young female's mating partners consist of all the males of that new mother-son unit (sons of the mother and the male who pairs with the mother). Even if we assume that mothers do not mate with their sons, this pattern smacks of sexual promiscuity. The offspring are the offspring of male-sibs or fathers, and are genetically highly related to each male of the unit. If the mother dies, the sons will probably separate. Some of them will form new pairs with the females who had already joined them before the mother's death. The other sons, failing this, will become solitary but sometime later will form new pairs.

In this way, the prototype of the human family may have taken the following course. First, a male-female pair was formed. Then, it developed into an expanded family including the original pair, their sons, and immigrant females. Finally, upon the mother's death, the family broke down into pairs and lone individuals.

Males and females employ different breeding strategies in the formation of pairs. For males, the energy cost required to pass genes to the next generation is only the discharge of sperm. Consequently, males do not select their mates and solicit several of them. For females, an increased rate of conception increases the probability that their genes will be passed to the next generation. In order for females to transmit their genes to the next generation, however, they must spend much time, and an enormous amount of energy, in pregnancy, delivery, and child-rearing. If there is a failure, en route, all of the females' endeavors will return nothing. If a female's egg unites with sperm containing non-adaptive genes, her breeding success becomes uncertain. Because the potential risks and costs are high, females, unlike males, select a mate.

One previously mentioned characteristic of chimpanzee society is that females transfer between groups, and these females are likely to select groups on the basis of their male members. Because our hominid ancestors perhaps shared with chimpanzees the characteristic of female selectivity and male nonselectivity, environmental pressures possibly were exerted and caused the first pair formation. Certainly, a prerequisite for pair formation is the continuous sexual receptivity of females.

Even granting, however, that continuous sexual receptivity has evolved in females, that alone does not account for the permanent pair. Because male sexual nonselectivity or promiscuity would have been as great an obstacle to the permanent pair, something in addition to continuous sexual activity would have had to have evolved to tie together specific individuals.

Looking at mother-offspring relations among the common chimpanzees at Gombe Park, we see that when a mother died, her offspring became lethargic. One individual displayed abnormal behavior, and another, while capable of feeding by himself, ate very little; the majority died (Goodall, 1971). The adaptive importance of attachment to the mother is understandable if, during infancy, the offspring have to depend completely on the mother to survive. At this age, the offspring may increase the protection and help they receive from their mothers through persistent attachment to them. After offspring become capable of subsisting on their own, however, dying from their mother's death would be maladaptive.

Even in adolescence and adulthood, the individual's continued maintenance of close relations with its mother may be explained by kin selection theory as the protection of family genes. Nevertheless, I wonder if attachment to the aged mother in humans, and perhaps even in pygmy chimpanzees, has any selective advantage that might explain this phenomenon in terms of sociobiological theory. An old female has already finished playing her role of delivering and caring for her children; her genes have been passed to the next generation. What can be the benefit of investing energy in her when she is old?

Attachment to the mother probably developed orthogenetically along with an increasingly prolonged maternal dependency that became progressively greater in each hominoid line. Some "overadaptation" might have occurred in this orthogenetic process, and immoderate depression over a mother's death or attachment to an aged mother may exemplify this.

One wonders if such a "love-attachment" to certain specified individuals, as well as a sexual attachment, was not a necessary or additional axis that supported the establishment of male-female pairs. The relative importance of these two axes changes from time to time. In the early formative period, the axis of sex is important. Later, that of "love" gradually becomes more important. Although this model includes the unreasonable assumption that the attachment to the mother is shifted to another female, if a "love-attachment" axis was necessary to establish ties between the male-female pair (in addition to sexual attachment), then the feeling called "love" could have been born.

In primate societies, the mother-offspring relationship is the only social structure that has increasingly developed friendly, mutual action over a history of 70 million years. Attachment to a specific individual must have originated in that structure. That is why I presented the hypothesis that the extended family composed of a mother, sons, and immigrant females is the original form of the human family. As a preliminary condition of pair formation, the attachment of offspring to the mother must be sufficiently established.

In both types of chimpanzee, females are the ones that separate from the mother. Even in the ancestors of humans, the attachment to a specific individual (the mother) may have been greater in males than in females. That probably promoted the male's positive attitude toward continuation of the pair.

Since the famous research concerning the pecking order of chickens (Schjelderup-Ebbe, 1922), Western researchers of animal behavior have tended to regard the rank system of hierarchical relations as the main plank supporting society. According to the Neo-Darwinian theory of intraspecific competition, a chicken of high pecking rank can send more genes to the next generation than a low-ranking individual. The high-ranking individual has priority in choices not only of food and sleeping places but also of mates. When rank is fixed, however, excess drive is controlled, and even a low-ranking chicken can secure some level of personal genetic survival. Dominant-subordinate relations within the same sex arise from sexual competition, and size, robustness, and intelligence are the sorts of traits that are selected in sexual competitions among males. Sexual differences appear in many primates (e.g., males are larger), as the result of sexual competition, usually among males.

Order is seen as being maintained by fixed hierarchical relations.

A comparison of the two chimpanzee species, however, seems to argue against individual coexistence arising from fixed hierarchical relations. In unit groups of common chimpanzees, a linear dominance hierarchy is established among mature males, and all adult males are dominant over all adult females. In pygmy chimpanzee unit groups also, dominant-subordinate relationships exist among adult males, but dominance interactions seldom occur in some dyads of adult males. Furthermore, dominance relationships between sexes are not unidirectional, as they are in common chimpanzees, and some females are dominant over some adult males. Thus, in pygmy chimpanzee society, the dominance hierarchy is relatively undeveloped and does not seem to operate as clearly as in common chimpanzee society.

If we compare these two species in terms of social phenomena such as group size, patterns of fission and fusion, intergroup aggression, and conspecific killing, we find that pygmy chimpanzees are more successful at individual coexistence than common chimpanzees. This implies that genuine coexistence will not be gained by the establishment of strict hierarchies but rather by the establishment of harmonious relationships based on equality among individuals. Among primates as a whole, however, there seems to be a contradictory tendency in that there are as many mutual aggressive actions as there are types of rank system. The establishment of rank serves to check the manifestation of "excessive" aggression but it is not something that fosters coexistence.

In general, dominance behavior and aggressive behavior in primates occur more often in males than in females. Likewise, although equal harmonious relations are maintained between pygmy chimpanzee males and females and between females, aggressive interactions occur much more frequently between males. In sociobiological theory, this would be an expression of sexual competition between males. If females are more selective in choosing a mate than males, it is strange that sexual competition is greater in males than in females.

Among pygmy chimpanzees, copulation is rarely interrupted by other males, and exclusive possession of estrous females by high-ranking males has not been observed. Among common chimpanzees at Mahale and Gombe, however, exclusive possession of females is common. Nevertheless, in pygmy chimpanzees the frequency of copulation is high in dominant males. Moreover, when a dominant individual is near, copulation by low-ranking individuals decreases. This positive correlation between rank and copulatory frequency, which appears to be intermale sexual competition, may actually result from female choice. This phenomenon merits further investigation.

At this point, I would like to propose another hypothesis concerning the high frequency of intermale aggression. This is not an alternative or opposing hypothesis, but rather a supplementary one filling a gap that cannot be encompassed by the hypothesis of sexual competition. We have noted that dominance or aggressive behavior in pygmy chimpanzees occurs at a higher frequency during feeding than in sexual contexts. In this case, dominance or aggressive behavior seems to relate to individual survival rather than to sexual competition. Why, then, do aggression and relationships of rank develop between males much more than between females?

The spatial arrangement of individual pygmy chimpanzees feeding at a vast food source, such as *Dialium* fruit, suggests an answer to this question. When pygmy chimpanzees arrive at such a food source, aggressive interactions frequently occur between males. After a while, the majority of males end up moving individually to other surrounding trees. The individuals remaining at the primary food source are females with their offspring and a few dominant males. Unlike the highly aggressive males, who must maintain a large individual space at feeding time, the females, who have a low level of mutual antagonism, can crowd together. Males have developed stronger time and spatial limits than females so that females and their offspring can use enough of the food base. The high level of intermale aggression serves the purpose of allowing the females, who are physically handicapped by having the burden of raising and carrying the infant, to take their time feeding.

This form of spatial arrangement becomes especially profitable for females and their offspring when they are using a large concentrated food resource. Many food resources are like that in the forest, and most forest primates form a uni-male type unit group, which is a social structure that can be regarded as an extreme pattern of individual spatial arrangement.

The hypothesis that the high level of intermale aggression is to help a female and her offspring to survive may apply to other primates that have multi-male unit groups. For example, in savanna baboons and macaques, the characteristic spatial pattern of individuals, in which females, the young, and small numbers of males gather in the center of the group and the majority of males scatter to the periphery, is maintained not only at feeding time but also during travel.

I do not intend to deny completely the role of sexual competition between males, but I think it occupies no more than a small part in explaining intermale aggression. Once the male-female pair bond evolved, enabling the male to provide economic assistance to the female and offspring, the large intermale distance would not be necessary for females and young to survive. That is, when the human family evolved, the need for so much male aggression disappeared.

The early hominids hunted and gathered food on the savanna. In the present human population, the hunting and gathering people of the Kalahari Desert practice the subsistence economy most similar to that adopted by the early hominids. In those people, there is no hierarchical ranking among males; the level of aggression is very low; and they form an "egalitarian" society (Itani, 1987). This example supports my supposition. If there had been any male aggression among hunting people in the millenia since the time of pair-bonding, it was selected out over a long period of time, and all that remains is the aspect of sexual competition.

Lorenz (1966) and Ardrey (1961) were wrong in thinking that human aggression was inherited long ago from hominoid ancestors. Instead, I agree with V. Reynolds (1967) that the aggressive and hierarchical behaviors found in existing human males appear to have originated secondarily, perhaps after the advent of farming. Farming is closely tied to sedentary habitation and increases the carrying capacity of the land. On the other hand, compared to a foraging subsistence, farming leads to the drying up of resources, and is subject to the influences of natural calamities. Because of these effects, the defense of one's territory becomes a vital matter. The budding of "territoriality" in human history may have occurred after the commencement of farming. In order to defend territory, fighting between neighbors became necessary. Moreover, farming, storage, and private property are tied together, and private property implies the need for defense. It also leads to the stratification of society and eventually to inequality among its members.

I suspect that aggression in human society developed after the emergence of agriculture, and that the aggression in primate males that developed mainly for the protection and care of females and their offspring was, originally, a completely different matter.

Appendixes

Wild Plant Foods of the Pygmy Chimpanzees of Wamba, 1975–81

(Kano and Mulavwa, 1984)

Plant	Part eaten	Life form	Principal habitat
Anacardiaceae			
Antrocaryon micraster A. Chev. and Cuillaum	Pulp	Tall Tree	PF
Annonaceae			
Anonidium mannii (Oliv.) Engl. and Diels	Pulp	Medium tree	PF
Polyalthia suaveolens Engl. and Diels	Pulp	Medium tree	PF
Apocynaceae			
Alstonia boonei De Wild.	Pulp	Low tree	PF, OSF
Ancylobotrys pyriforms Pierre	Pulp	Woody vine	PF
Baissea thollonii Hua.	Pulp	Woody vine	PF, OSF
Dictyophleba ochracea (K. Schum. ex Hallier f.) Pichon	Pulp	Woody vine	PF
Landolphia angolensis (Stapf) A. Rich	Pulp	Woody vine	OSF, PF
Landolphia congolensis (Stapf) Pichon	Pulp	Woody vine	OSF
Landolphia ligustrifolia (Stapf) M. Pichon	Pulp	Woody vine	PF
Landolphia owariensis P. Beauv.	Pulp	Woody vine	PF, OSF
Landolphia violacea (K. Schum. ex Hallier f.) Pichon	Pulp	Woody vine	PF
Burseraceae			
Canarium schweinfurthii Engl.	Pulp	Tall tree	PF
Dacryodes edulis (G. Don) H. J. Lam.	Pulp	Tall tree	PF
Santiria trimera (Oliv.) Aubrev.	Pulp	Tall tree	PF

Unidentified food types: leaf (five species), fruit (four species), mushrooms (two species), dead branch (one species), and petiole (one species).
PF, Primary forest; OSF, old secondary forest; YSF, young secondary forest; SWF, swamp forest; SB, secondary bush

Plant	Part eaten	Life form	Principal habitat
Caesalpiniaceae			
Afzelia bipindensis Harms	Leaf, pod, seed	Tall tree	PF
Brachystegia laurentii (De Wild.) Louis	Seed	Tall tree	PF
Crudia harmsiana De Wild.	Seed	Tall tree	PF
Cynometra alexandri C. H. Wright	Pod, seed	Tall tree	PF
Cynometra hankei Harms	Leaf	Tall tree	PF
Dialium corbisieri Staner	Pulp, leaf	Tall tree	PF
Dialium excelsum Louis ex Steyaert	Pulp, young leaf	Tall tree	PF
Dialium pachyphyllum Harms	Pulp	Tall tree	PF
Dialium zenkeri Harms	Pulp	Tall tree	PF
Erythrophleum suaveoleus (Guill. and Perr.) J. Brenan	Seed	Tall tree	PF
Gilbertiodendron dewevrei (De Wild.) J. Léonard	Seed	Tall tree	PF
Leonardoxa romii (De Wild.) Aubrev.	Young leaf, stem	Medium tree	PF
Anthonotha macrophylla De Wild.	Seed	Low tree	PF
Anthonotha fragrans (Bak. f.)	Seed	Tall tree	PF
Scorodophloeus zenkeri Harms	Seed, leaf, flower	Tall tree	PF, OSF
Capparidaceae			
Pentadiplandra brazzeana Baill.	Pulp	Vine	YSF
Commelinaceae			
Palisota ambigua (P. Beauv.) C. B. Cl.	Herbaceous stem	Herb	PF, OSF, SWF
Palisota brachythyrsa Mildbr.	Herbaceous stem	Herb	PF, OSF, YSF
Palisota schweinfurthii C. B. Cl.	Herbaceous stem	Herb	PF, OSF, SWF
Ebenaceae			
Diospyros alboflavescens F. White	Pulp	Low tree	PF, OSF
Euphorbiaceae			
Croton haumanianus J. Léonard	Pulp	Tall tree	YSF
Dichostemma glaucescens Pierre	Pulp	Low tree	SWF
Drypetes angustifolia Pax and K. Hoffm.	Pulp	Tall tree	PF
Drypetes louisii J. Léonard	Pulp	Tall tree	PF
Macaranga spinosa Müll Arg.	Leaf	Tall tree	YSF
Manniophyton fulvum Müll. Arg.	Leaf	Woody vine	PF, OSF
Uapaca guineensis Müll Arg.	Pulp, petiole	Tall tree	OSF
Uapaca heudelotii Baill.	Pulp	Tall tree	SWF

Appendixes

Plant	Part eaten	Life form	Principal habitat
Flacourtiaceae			
Caloncoba welwitschii (Oliv.) Gilg	Pulp	Low tree	YSF, SB
Guttiferae			
Mammea africana Sabine	Pulp	Medium tree	PF
Symphonia globulifera L. f.	Pulp	Tall tree	SWF
Hippocrateaceae			
Cuervea macrophylla (Vahl) R. Wilczek ex Hallé	Seed, leaf	Woody vine	PF
Salacia alata De Wild.	Seed	Vine or herb	PF
Irvingiaceae			
Irvingia gabonensis (Aubry Lecomte ex O'Rorke) Baill.	Pulp	Tall tree	PF
Ixonanthaceae			
Ochthocosmum africanus Hook f.	Leaf	Tall tree	PF, OSF
Lauraceae			
Beilschmiedia corbisieri (Robyns) Robyns and Wilczek	Pulp	Tall tree	PF
Beilschmiedia variabilis Robyns and Wilczek	Pulp	Tall tree	PF
Malvaceae			
Sida rhombifolia L.	Pulp	Shrub	YSF
Marantaceae			
Ataenidia conferta (Benth.) Milne-Redh.	Shoot, pulp	Herb	PF, SWF
Haumania liebrechtsiana (De Wild. and Th. Dur.) J. Leon	Shoot, petiole	Woody vine	PF, OSF
Hypselodelphys poggeana (K. Schum.) Milne-Redh.	Shoot, petiole	Woody vine	PF, OSF
Hypselodelphys scandens Louis and Mullend.	Shoot, petiole	Woody vine	OSF
Hypselodelphys violacea (Ridl.) Milne-Redh.	Shoot, petiole	Woody vine	PF, OSF
Megaphrynium macrostachyum (Benth.) Milne-Redh.	Shoot, pulp	Herb	OSF
Trachyphrynium braunianum (K. Schum.) Baker	Pulp	Herb	SB
Melastomataceae			
Memecylon jasminoides Gilg.	Pulp	Tall tree	PF
Menispermaceae			
Triclisia dictyophylla Diels	Pulp	Woody vine	PF
Mimosaceae			
Albizzia gummifera C. A. Sm. var. *ealaënsis* (De Wild.) Brenan	Leaf	Tall tree	YSF
Pentaclethra macrophylla Benth.	Seed, bark	Tall tree	PF

Plant	Part eaten	Life form	Principal habitat
Moraceae			
Antiaris welwitschii Engl.	Pulp	Tall tree	PF
Ficus ottoniifolia (Miq.) Miq.	Pulp	Tall tree (woody vine)	PF
Musanga smithii B. Br.	Pulp	Medium tree	YSF
Myrianthus arboreus P. Beauv.	Pulp	Medium tree	OSF, YSF
Treculia africana Dence	Pulp	Tall tree	PF
Olacaceae			
Ongokea gore (Hua) Pierre	Pulp	Tall tree	PF
Palmae			
Ancistrophyllum secundiflorum (P. Beauv.) Wendl.	Pith	Woody vine	OSF
Eremospatha hookeri (Mann and Wendl.) Wendl.	Shoot	Woody vine	PF
Sclerosperma mannii Wendl.	Shoot	Low tree	SWF
Papilionaceae			
Baphia laurifolia Baill.	Leaf	Tall tree	PF
Dalbergia lactea Vatke	Leaf	Tall tree	YSF
Millettia duchesnei De Wild.	Seed	Tall tree	PF
Pterocarpus casteelsii De Wild.	Young leaf	Tall tree	PF
Piperaceae			
Piper guineense Schum. and Thonn.	Pulp	Woody vine	PF
Polygalaceae			
Carpolobia alba G. Don	Pulp	Low tree	PF, OSF
Rosaceae			
Parinari excelsa Sabine	Pulp	Tall tree	PF
Sapindaceae			
Allophylus africanus P. Beauv.	Pulp	Low tree	YSF
Chytranthus carneus Radlk. ex Mildbr.	Young leaf	Low tree	PF
Pancovia laurentii (De Wild.) Gilg ex De Wild.	Pulp	Medium tree	PF, OSF
Sapotaceae			
Manilkara casteelsii (De Wild.) Evrard	Pulp	Tall tree	PF
Pachystela excelsa [au. not given]	Pulp	Tall tree	PF
Synsepalum subcordatum De Wild.	Pulp	Tall tree	PF, OSF
Sterculiaceae			
Cola bruneelii De Wild.	Leaf	Shrub	PF, OSF
Cola griseiflora De Wild.	Pulp	Medium tree	PF
Cola marsupium K. Schum.	Pulp, leaf	Low tree	OSF
Leptonychia batangensis (C. H. Wright) Burret	Pulp	Low tree	PF
Leptonychia tokana R. Germain	Pulp	Low tree	PF
Sterculia bequaertii De Wild.	Leaf	Tall tree	PF

Plant	Part eaten	Life form	Principal habitat
Tiliaceae			
Grewia malacocarpa Mast.	Pulp	Tall tree	PF
Grewia pinnatifida Mast.	Pulp, young leaf	Medium tree	PF, OSF
Vitaceae			
Cissus aralioides (Welw. ex Bak.) Planch.	Pulp	Woody vine	OSF, PF
Cissus dasyantha Gilg and Brandt	Pulp	Woody vine	PF, OSF
Cissus dinklagei Gilg and Brandt	Pulp	Woody vine	PF, OSF
Zingiberaceae			
Aframomum laurentii De Wild.	Pith	Herb	SWF
Aframomum sp.	Pulp, pith	Herb	SB
Renealmia africana Benth. ex Hook. F.	Pith	Herb	SWF
Fungus			
?Langermania fenzlii [au. not given]	Whole	Mushroom	PF

Cultigens and Other Foods of the Pygmy Chimpanzees of Wamba

Cultivated plants

Pineapple (pulp)
Sugarcane (stalk)
Papaya (pulp; provided by humans, only one observed case)
Banana (pith)
Coffee (pulp; only one case of food traces)

Animal or other products

Flying squirrel (*Anomalurus erythrotus*)
Earthworm (two species)
Butterfly larvae (*Andronymus neander* Plötz, Hesperidae)
Honey of stingless bees (*Apidae* sp. and *Dactyfurina staudingeri* Gribodo)
Soil from the nests of termites (*Cubitermes* sp.)
Chicken eggs (provided by humans)

Wild Plant Foods of the Pygmy Chimpanzees of Yalosidi

Plant	Part eaten	Life form	Principal habitat
Alismataceae			
Ranalisma humile (Kunth.) Hutch.	Leaf, stem, root	Herb	SWG
Anacardiaceae			
Trichoscypha ferruginea Engl.	Pulp	Tall tree	PF
Annonaceae			
Anonidium mannii (Oliv.) Engl. and Diels	Pulp	Medium tree	PF
Cleistopholis glauca Pierre ex Engl. and Diels	Pulp	Tall tree	YSF
Apocynaceae			
Landolphia owariensis P. Beauv.	Pulp	Woody vine	PF
Rauvolfia vomitoria Afzel	Pulp	Medium tree	OSF
Caesalpiniaceae			
Dialium corbisieri Staner	Pulp	Tall tree	PF
Commelinaceae			
Palisota ambigua (P. Beauv.) C. B. Cl.	Stem	Herb	OSF, PF
Cyperaceae			
Cyperus nudicaules [au. not given]	Base of stem	Herb	SWG
Euphorbiaceae			
Croton haumanianus J. Léonard	Pulp	Tall tree	YSF
Drypetes cinnabarina Pax and K. Hoffm.	Pulp	Medium tree	YSF
Spondianthus preussii Engl.	Pulp	Tall tree	OSF, PF
Uapaca guineensis Mull. Arg.	Pulp	Tall tree	YSF, SWF
Guttiferae			
Garcinia pynaertii De Wild.	Pulp	Tall tree	SWF

PF, primary forest; OSF, old secondary forest; YSF, young secondary forest; SWF, swamp forest; SWG, swamp grassland

Plant	Part eaten	Life form	Principal habitat
Irvingiaceae			
Irvingia gabonensis (Aubry Lecomte ex O'Rorke)	Pulp	Tall tree	PF
Marantaceae			
Haumania liebrechtsiana (De Wild. ex Th. Dur.) J. Leon	Shoot, stem	Herb (vine)	YSF, OSF
Marantochloa congensis (K. Schum.) Léonard and Mull.	Shoot	Herb	SWG
Sarcophyrynium macrostachyum (Benth.) Milne-Redh.	Shoot, pulp, leaf, stem	Herb	YSF, OSF
Mimosaceae			
Albizzia gummifera C. A. Sm. var. *ealaensis* (De Wild.)	Pulp	Tall tree	YSF
Moraceae			
Musanga cecropioides R. Br.	Pulp	Tall tree	YSF
Treculia africana Dence	Pulp	Tall tree	PF
Olacaceae			
Strombosiopsis tetrandra Engl. J. Brenan	Pulp	Low tree	YSF
Palmae			
Ancistrophyllum secundiflorum (P. Beauv.) Wendl.	Pith	Woody vine	OSF, YSF
Eremospatha haullevileana De Wild.	Shoot	Low tree	SWF
Papilionaceae			
Millettia drastica Welw. ex Bak.	Pulp	Medium tree	YSF
Rosaceae			
Parinari excelsa Sabine	Pulp	Tall tree	YSF, OSF
Sapotaceae			
Gambeya lacourtiana Auba. and Pellergr.	Pulp	Tall tree	OSF, PF
Omphalocarpum lecomteanum Pierre ex Engl.	Pulp	Tall tree	PF
Sterculiaceae			
Cola bruneelii De Wild.	Leaf	Medium tree	PF
Tiliaceae			
Grewia coriaceae Mast.	Pulp	Medium tree	OSF
Violaceae			
Rinorea oblongifolia Marquand ex Chipp.	Pulp	Tall tree	OSF, PF
Vitaceae			
Cissus leonardii Dewit.	Pulp	Woody vine	OSF, PF
Zingiberaceae			
Aframomum laurentii De Wild.	Pulp	Herb	SWG
Aframomum sp.	Pith, stem, pulp	Herb	YSF
Renealmia africana Benth. ex Hook. f.	Stem	Herb	SWF

References Cited
and Index

References Cited

Agar, M. H. 1980. *The Professional Stranger*. Orlando: Academic Press.

Andrews, P., and J. E. Cronin. 1982. The relationship of *Sivapithecus* and *Ramapithecus* and the evolution of the orang-utan. *Nature* 297: 541–46.

Ardrey, R. 1961. *African Genesis*. London: Collins.

———. 1966. *The Territorial Imperative*. London: Collins.

———. 1970. *The Social Contract*. London: Collins.

Asquith, P. J. 1981. Some aspects of anthropomorphism in the terminology and philosophy underlying Western and Japanese studies of the social behaviour of non-human primates. D. Phil. thesis, University of Oxford.

———. 1986. Anthropomorphism and the Japanese and Western traditions in primatology. In: *Primate Ontogeny, Cognition, and Social Behaviour*, pp. 61–71. J. G. Else and P. C. Lee, eds. Cambridge, Eng.: Cambridge University Press.

Badrian, A., and N. Badrian. 1977. Pygmy chimpanzees. *Oryx* 13: 463–68.

———. 1984. Social organization of *Pan paniscus* in the Lomako forest, Zaire. In: *The Pygmy Chimpanzee: Evolutionary Biology and Behavior*, pp. 275–79. R. L. Susman, ed. New York: Plenum Press.

Badrian, N., A. Badrian, and R. L. Susman. 1981. Preliminary observations on the feeding behavior of *Pan paniscus* in the Lomako forest of central Zaire. *Primates* 22(2): 173–81.

Badrian, N., and R. K. Malenky. 1984. Feeding ecology of *Pan paniscus* in the Lomako forest, Zaire. In: *The Pygmy Chimpanzee: Evolutionary Biology and Behavior*, pp. 275–99. R. L. Susman, ed. New York: Plenum Press.

Barzun, J. 1986. *On Writing, Editing and Publishing*. 2d ed. Chicago: University of Chicago Press.

Barzun, J., and H. F. Graff. 1977. *The Modern Researcher*. New York: Harcourt Brace Jovanovich.

Battel, A. 1625. In: Purchas, S. 1925. *Hakluytus Posthumaus, or Purchas His Pilgrimes. Contayning a History of the World, in Sea Voyages & Lande Travell*, 4th ed., Vol. 2. London.

Bernstein, I. S. 1969. A comparison of nesting patterns among the three great apes. In: *The Chimpanzee*, Vol. 1, pp. 393–402. G. H. Bourne, ed. Basel: Karger.

Bourne, G. H. 1949. Vitamin C and immunity. *Brit. J. Nutr.* 2: 341–47.

———. 1971. Nutrition and diet of chimpanzees. In: *The Chimpanzee*, Vol. 4, pp. 373–400. G. H. Bourne, ed. Basel: Karger.

Bournonville, D. de. 1967. Contribution a l'étude du chimpanzé de Guinée. *Bull. Inst. Fr. Afr. Noire* 29: 1189–1269.

Bree, P. J. H. van. 1963. On a specimen of *Pan paniscus* Schwarz, 1929, which lived in the Amsterdam Zoo from 1911 till 1916. *Zool. Garten N.F.* 17: 292–95.

Broom, R. 1925. On the newly discovered South African man-ape. *Nat. Hist.* 25: 409–18.

———. 1936. A new fossil anthropoid skull from South Africa. *Nature* 138: 486–88.

Callewaert, T. 1930. Les chimpanzés de la rive gauche du Congo. *Bull. Cercl. zool. congol.* 6: 67–69.

Carpenter, C. R. 1942. Sexual behavior of free ranging rhesus monkeys. *J. Comp. Psych.* 33: 113–62.

Carroll, R. 1988. *Cultural Misunderstandings: The French-American Experience*, translated by Carol Volk. Chicago: University of Chicago Press.

Coolidge, H. J. 1933. *Pan paniscus*: Pygmy chimpanzee from south of the Congo River. *Am. J. Phys. Anthropol.* 18(1): 1–57.

Cramer, D. L. 1977. Craniofacial morphology of *Pan paniscus*: A morphometric and evolutionary appraisal. *Contrib. Primatol.* 10: 1–64.

Dart, R. A. 1925. *Australopithecus africanus*: The man-ape of South Africa. *Nature* 115(2884): 195–99.

Darwin, C. 1871. *The Descent of Man and Selection in Relation to Sex*. London: Murray.

DeBeaugrande, R. 1984. *Text Production* (Advances in Discourse Processes, Vol. 11). Norwood, Mass.: Ablex.

Elliot, D. G. 1913. A review of the primates. *Monogr. No. 1. Am. Mus. Nat. Hist.*, Vol. 3.

Elliot, R. C. 1976. Observations on a small group of mountain gorillas (*Gorilla gorilla beringei*). *Folia primatol.* 25: 12–24.

Evrard, C. 1968. Recherches écologiques sur le peuplement forestier des sols hydromorphes de la cuvette centrale congolaise. *I.N.E.A.C. série scientifique* 110: 1–295.

Fobes, J. L., and J. E. King. 1982. *Primate Behavior*. New York: Academic Press.

Frisch, J. 1959. Research on primate behavior in Japan. *Amer. Anthro.* 61(4): 584–96.

———. 1963. Japan's contribution to modern anthropology. In: *Studies in Japanese Culture*, pp. 225–44. J. Roggendorf, ed. Tokyo: Sophia University.

Galdikas, B. M. F. 1979. Orangutan adaptation at Tanjung Puting Reserve:

Mating and ecology. In: *The Great Apes*, pp. 192–233. D. A. Hamburg and E. R. McCown, eds. Menlo Park, CA: Benjamin/Cummings.

———. 1981. Orangutan reproduction in the wild. In: *Reproductive Biology of the Great Apes: Comparative and Biomedical Perspectives*, pp. 281–99. C. E. Graham, ed. New York: Academic Press.

Gijzen, A. 1974. Studbook of *Pan paniscus* Schwarz, 1929. *Acta zool. pathol. Antverp.* 61: 119–64.

Goodall, A. G. 1977. Feeding and ranging behaviour of a mountain gorilla group (*Gorilla gorilla beringei*) in the Tshibinda-Kahuzi Region (Zaire). In: *Primate Ecology*, pp. 450–79. T. S. Clutton-Brock, ed. London: Academic Press.

Goodall, J. 1965. Chimpanzees of the Gombe Stream Reserve. In: *Primate Behavior: Field Studies of Monkeys and Apes*, pp. 425–73. I. DeVore, ed. New York: Holt, Rinehart, and Winston.

———. 1968. The behaviour of free-living chimpanzees of the Gombe Stream Reserve. *Anim. Behav. Monogr.* 1: 161–311.

———. 1970. Tool-using in primates and other vertebrates. In: *Advances in the Study of Behavior*, Vol. 3, pp. 195–249. D. S. Lehrman, R. S. Hinde, and E. Shaw, eds. New York: Academic Press.

———. 1971. *In the Shadow of Man*. London: Collins.

———. 1977. Infant killing and cannibalism in free-living chimpanzees. *Folia primatol.* 28: 259–82.

———. 1983. Population dynamics during a 15-year period in one community of free-living chimpanzees in the Gombe National Park, Tanzania. *Z. Tierpsychol.* 61: 1–60.

Goodall, J., A. Bandora, E. Bergman, C. Busse, H. Matama, E. Mpongo, A. Pierce, and D. Riss. 1979. Intercommunity interactions in the chimpanzee population of the Gombe National Park. In: *The Great Apes*, pp. 13–53. D. A. Hamburg, and E. R. McCown, eds. Menlo Park, CA: Benjamin/Cummings.

Grant, C. H. B. 1946. The distribution of the chimpanzee in Tanganyika Territory. *Tanganyika Notes Rec.* 21: 110–111.

Gribbin, J., and J. Cherfas. 1981. Descent of man or ascent of ape? *New Scientist* 91: 592–95.

Hamilton, A. 1976. The significance of patterns of distribution shown by tropical plants and animals in tropical Africa for the reconstruction of upper Pleistocene palaeoenvironments: A review. *Paleoecol. Afr.* 9: 63–97.

Handler, N. T., R. K. Malenky, and N. Badrian. 1984. Sexual behavior of *Pan paniscus* under natural conditions in the Lomako forest, Equateur, Zaire. In: *The Pygmy Chimpanzee: Evolutionary Biology and Behavior*, pp. 347–68. R. L. Susman, ed. New York: Plenum Press.

Harcourt, A. H., and K. J. Stewart. 1978. Sexual behavior of wild gorillas. In: *Recent Advances in Primatology*, Vol. 1, *Behaviour*, pp. 611–12. D. J. Chivers, and J. Herbert, eds. New York: Academic Press.

Harcourt, A. H., K. J. Stewart, and D. Fossey. 1981. Gorilla reproduction in the wild. In: *Reproductive Biology of the Great Apes: Comparative and Biomedical Perspectives*, pp. 265–78. C. E. Graham, ed. New York: Academic Press.

Hasegawa, M. 1984. DNA kara mita jinrui no kigen to shinka—bunshi jinruigaku yosetsu. [The origin and evolution of man based on DNA analysis: An introduction to molecular anthropology.] Tokyo: Kaimeisha.

Hess, J. P. 1973. Some observations on the sexual behaviour of captive lowland gorillas. In: *Comparative Ecology and Behaviour of Primates*, pp. 507–581. R. P. Michael, and J. H. Crook, eds. London: Academic Press.

Hladik, C. M. 1977. Chimpanzees of Gabon and chimpanzees of Gombe: Some comparative data on the diet. In: *Primate Ecology*, pp. 481–501. T. H. Clutton-Brock, ed. London: Academic Press.

Horn, A. D. 1979. The taxonomic status of the bonobo chimpanzee. *Am. J. Phys. Anthropol.* 51: 273–82.

———. 1980. Some observations on the ecology of the bonobo chimpanzee (*Pan paniscus* Schwarz, 1929) near Lake Tumba, Zaire. *Folia primatol.* 34: 145–69.

Itani, J. 1972. *Reichourui no shakai kouzou.* (Social structure of primates.) Tokyo: Kyoritsu Shuppan.

———. 1979. Distribution and adaptation of chimpanzees in an arid area. In: *The Great Apes*, pp. 123–35. D. A. Hamburg, and E. R. McCown, eds. Menlo Park, Calif.: Benjamin/Cummings.

———. 1987. Inequality versus equality for coexistence in primate societies. In: *Dominance, Aggression and War*, pp. 75–104. D. McGuinness, ed. New York: Paragon House.

Janzen, D. H. 1975. *Ecology of Plants in the Tropics.* London: Edward Arnold.

Johanson, D. C. 1974. Some metric aspects of the permanent and deciduous dentition of the pygmy chimpanzee (*Pan paniscus*). *Am. J. Phys. Anthropol.* 41: 39–48.

Johanson, D. C., and T. D. White. 1979. A systematic assessment of early African hominids. *Science* 203(4378): 321–30.

Johnson, S. C. 1981. Bonobos: Generalized hominid prototypes or specialized insular dwarfs? *Current Anthro.* 22(4): 363–75.

Johnston, H. 1922. 'Introduction' to T. A. Barns' *The Wonderland of the Eastern Congo*. London.

Jones, C., and J. Sabater-Pi. 1971. Comparative ecology of *Gorilla gorilla* (Savage and Wymann) and *Pan troglodytes* (Blumenbach) in Rio Muni, West Africa. *Biblio. Primat.* 13: 1–96.

Kadomura, H. 1980. Reconstruction of palaeoenvironments in Tropical Africa during the Wurm glacial maximum: A review. *Environ. Sci., Hokkaido* 3(2): 147–54.

Kano, T. 1972. Distribution and adaptation of the chimpanzee on the eastern shore of Lake Tanganyika. *Kyoto Univ. Afr. Studies* 7: 37–129.

———. 1982a. Social group of pygmy chimpanzees (*Pan paniscus*) of Wamba. *Primates* 23: 171–88.

————. 1982b. The use of the leafy twigs for rain cover by the pygmy chimpanzees of Wamba. *Primates* 23(3): 453–57.

————. 1983. An ecological study of the pygmy chimpanzees (*Pan paniscus*) of Yalosidi, Republic of Zaire. *Int. J. Primat.* 4: 1–31.

————. 1984a. Distribution of pygmy chimpanzees (*Pan paniscus*) in the central Zaire basin. *Folia primatol.* 43: 36–52.

————. 1984b. Observations of physical abnormalities among the wild bonobos (*Pan paniscus*) of Wamba, Zaire. *Am. J. Phys. Anthropol.* 63: 1–11.

Kano, T., and M. Mulavwa. 1984. Feeding ecology of the pygmy chimpanzee (*Pan paniscus*) of Wamba. In: *The Pygmy Chimpanzee: Evolutionary Morphology and Behavior*, pp. 233–74. R. L. Susman, ed. New York: Plenum.

Kawai, M. 1979. Shinrin ga saru o sunda—genzai no shizensho. Tokyo: Heibonsha.

Kawanaka, K. 1979. Age-sex composition and size of M-group of Kasoje. *Mahali Mountains Chimpanzee Res. Rep.* 5.

————. 1981. Infanticide and cannibalism in chimpanzees, with special reference to the newly observed case in the Mahale mountains. *Kyoto Univ. Afr. Study Monogr.* 1: 69–99.

Kindaichi, H. 1978. *The Japanese Language*, translated by U. Hirano. Rutland, Vt.: Charles E. Tuttle.

King, M.-C., and A. C. Wilson. 1975. Evolution at two levels in humans and chimpanzees. *Science* 188(331): 107–16.

Kinzey, W. G. 1984. The dentition of the pygmy chimpanzee, *Pan paniscus*. In: *The Pygmy Chimpanzee: Evolutionary Biology and Behavior*, pp. 65–88. R. L. Susman, ed. New York: Plenum Press.

Kline, L. W. 1978. That which is lost in translation. In: *Cross-Cultural Perspectives on Reading and Reading Research*, pp. 191–201. Newark: International Reading Association.

Kollar, E. J., W. C. Beckwith, and R. B. Edgerton. 1968. Sexual behavior of the ARL Colony chimpanzees. *J. Nerv. Ment. Disease* 147(5): 444–59.

Kortlandt, A. 1962. Chimpanzees in the wild. *Sci. Am.* 206(5): 128–38.

————. 1967. Handgebrauch bei freilebenden Schimpansen. In: *Handgebrauch und Verstandigung*, pp. 59–102. Affen and Fruhmenschen, ed.

————. 1983. Marginal habitats of chimpanzees. *J. Human Evol.* 12: 231–78.

Koyama, N. 1977. Nihonzaru no shakai kouzou. [The social structure of Japanese macaques.] In: *Anthropology 2: Primates*, pp. 225–76. Editorial Board of Anthropology, ed. Tokyo: Yuzankaku.

Kuroda, S. 1984. Interaction over food among pygmy chimpanzees. In: *The Pygmy Chimpanzee: Evolutionary Biology and Behavior*, pp. 301–24. R. L. Susman, ed. New York: Plenum Press.

Lancaster, J. B., and R. B. Lee. 1965. The annual reproductive cycle in monkeys and apes. In: *Primate Behavior: Field Studies of Monkeys and Apes*, pp. 486–513. I. DeVore, ed. New York: Holt, Rinehart, and Winston.

Latimer, B. M., T. D. White, W. H. Kimbel, and D. C. Johanson. 1981. The pygmy chimpanzee is not a living missing link in human evolution. *J. Human Evol.* 10: 475–88.

Lebra, T. S., and W. P. Lebra, eds. 1986. *Japanese Culture and Behavior, Selected Readings*. Rev. ed. Honolulu: University of Hawaii Press.

Lemmon, W. B., and M. L. Allen. 1978. Continual sexual receptivity in the female chimpanzee (*Pan troglodytes*). *Folia primatol.* 30: 80–88.

Lorenz, K. 1966. *On Aggression*. London: Methuen.

MacKinnon, J. 1979. Reproductive behavior in wild orangutan populations. In: *The Great Apes*, pp. 257–73. D. A. Hamburg, and E. R. McCown, eds. Menlo Park, CA: Benjamin/Cummings.

McGrew, W. C., P. J. Baldwin, and C. E. G. Tutin. 1981. Chimpanzees in a hot, dry, and open habitat: Mt. Assirik, Senegal, West Africa. *J. Human Evol.* 10: 227–44.

McKenna, J. J. 1982. The evolution of primate societies, reproduction, and parenting. In: *Primate Behavior*, pp. 87–133. J. L. Fobes, and J. E. King, eds. New York: Academic Press.

———. 1983. Primate aggression and evolution: An overview of sociobiological and anthropological perspectives. *Bull. Am. Acad. Psychiatry Law* 11(3): 105–30.

Marcus, G. E., and M. J. Fischer. 1986. *Anthropology and Cultural Critique*. Chicago: University of Chicago Press.

Martlew, M. 1983. Problems and difficulties: Cognitive and communication aspects of writing. In: *The Psychology of Written Language*, pp. 295–333. M. Martlew, ed. Chichester: John Wiley.

Matsumoto, M. 1988. *The Unspoken Way*. Tokyo: Kodansha International.

Miller, L. C., and R. D. Nadler. 1981. Mother-infant relations and infant development in captive chimpanzees and orang-utans. *Int. J. Primat.* 2(3): 247–61.

Miller, R. A. 1977. *The Japanese Language in Contemporary Japan: Some sociolinguistic observations*. Washington, D.C.: American Enterprise Institute for Public Policy Research.

———. 1986. *Nihongo: In defence of Japanese*. London: Athlone Press.

Nadler, R. D. 1975. Sexual cyclicity in captive lowland gorillas. *Science* 189: 813–14.

———. 1977. Sexual behavior of the chimpanzee in relation to the gorilla and orang-utan. In: *Progress in Ape Research*, pp. 191–206. G. H. Bourne, ed. New York: Academic Press.

Nakane, C. 1972. *Japanese Society*. Berkeley: University of California Press.

Napier, J. R., and P. H. Napier. 1967. *A Handbook of Living Primates*. London: Academic Press.

Nishida, T. 1968. The social group of wild chimpanzees in the Mahali Mountains. *Primates* 9: 167–224.

———. 1970. Social behavior and relationship among wild chimpanzees of the Mahali Mountains. *Primates* 11: 47–87.

———. 1972. Preliminary information of the pygmy chimpanzees (*Pan paniscus*) of the Congo basin. *Primates* 13: 415–25.

————. 1974. Yasei chimpanji no seitai. (The ecology of wild chimpanzees.) In: *Jinrui no seitai*, pp. 15–60. J. Tanaka, and T. Nishida, eds. Tokyo: Kyoritsu Shuppan.

————. 1977. Mahare san no chimpanji (1) Seitai to tan'i shudan no kouzou (The chimpanzees of Mahale Mts. (1) Ecology and structure of the unit group.) In: *Chimpanji-ki*, pp. 543–638. J. Itani, ed. Tokyo: Kodansha.

————. 1979a. The social structure of chimpanzees of the Mahali Mountains. In: *The Great Apes: Perspectives on Human Evolution*, pp. 73–121. D. Hamburg, and E. McCown, eds. Menlo Park, CA: Benjamin/Cummings.

————. 1979b. The leaf-clipping display: A newly discovered expressive gesture in wild chimpanzees. *J. Human Evol.* 9: 117–28.

Nishida, T., M. Hasegawa, T. Hasegawa, and Y. Takahata. 1985. Group extinction and female transfer in wild chimpanzees in the Mahale National Park, Tanzania. *Z. Tierpsychol.* 67: 284–301.

Nishida, T., and M. Hiraiwa. 1982. Natural history of a tool-using behavior by wild chimpanzees in feeding upon wood-boring ants. *J. Human Evol.* 11: 73–99.

Nishida, T., and K. Kawanaka. 1972. Inter-unit group relationships among wild chimpanzees of the Mahali Mountains. *Kyoto Univ. Afr. Studies* 7: 131–69.

Patterson, T. 1979. The behavior of a group of captive pygmy chimpanzees (*Pan paniscus*). *Primates* 20(3): 341–54.

Petersen, G. B. 1979. *The Moon in the Water: Understanding Tanizaki, Kawabata, and Mishima*. Honolulu: University Press of Hawaii.

Portielje, A. F. J. 1916. *Een gids bij uw rondgang*. Amsterdam: Artis-gids.

Purves, A. C., ed. 1988. *Writing Across Languages and Cultures: Issues in Contrastive Rhetoric*. Newbury Park, Calif.: SAGE.

Radice, W., and B. Reynolds, eds. 1987. *The Translator's Art: Essays in Honour of Betty Radice*. Middlesex: Penguin.

Rahm, U. 1967. Observations during chimpanzee captures in the Congo. In: *Neue Ergebnisse der Primatologie*, pp. 195–207. D. Strack, R. Schneider, and H. J. Kuhn, eds. Stuttgart: Gustav Fischer.

Reynolds, V. 1966. Open group in hominid evolution. *Man* 1: 441–52.

————. 1967. *The Great Apes*. London: Cussel.

Reynolds, V., and F. Reynolds. 1965. Chimpanzees of the Budongo Forest. In: *Primate Behavior: Field Studies of Monkeys and Apes*, pp. 368–424. I. DeVore, ed. New York: Holt, Rinehart, and Winston.

Rijksen, H. D. 1978. *A Field Study on Sumatran Orang Utans (Pongo pygmaeus abelii Lesson 1827): Ecology, Behaviour and Conservation*. Wageningen, Netherlands: H. Veenman and B. V. Zonen.

Riss, D., and J. Goodall. 1977. The recent rise to the alpha-rank in population of free-living chimpanzees. *Folia primatol.* 27: 134–51.

Sarich, V. M. 1984. Pygmy chimpanzee systematics: A molecular perspective. In: *The Pygmy Chimpanzee: Evolutionary Biology and Behavior*, pp. 43–48. R. L. Susman, ed. New York: Plenum Press.

Sarich, V. M., and A. C. Wilson. 1967. Immunological time scale for hominid evolution. *Science* 154: 1200–1203.

Savage-Rumbaugh, E. S., and B. J. Wilkerson. 1978. Socio-sexual behavior in *Pan paniscus* and *Pan troglodytes*: A comparative study. *J. Human Evol.* 7: 327–44.

Schaller, G. B. 1965. The behaviour of the mountain gorilla. In: *Primate Behavior: Field Studies of Monkeys and Apes*, pp. 324–67. I. DeVore, ed. New York: Holt, Rinehart, and Winston.

Schjelderup-Ebbe, T. 1922. Beitrage zur Sozialpsychologie des Haushuhns. *Z. psychol.* 88: 225–52.

Schouteden, H. 1928. Le chimpanzé du Congo central. *Bull. Cercl. zool. congol.* 6: 108.

———. 1931. Quelques notes sur le chimpanzé de la rive gauche du Congo, *Pan satyrus paniscus*. *Rev. Zool. Bot. afr.* 20: 310–14.

Schultz, A. H. 1956. The occurrence and frequency of pathological and teratological conditions and of twinning among non-human primates. *Primatologia* 1: 965–1014.

———. 1966. Changing views on the nature and interactions of the higher primates. *Yerkes News* 3(1): 15–29.

Schwarz, E. 1928. Le chimpanzé de la rive gauche du Congo. *Bull. Cercl. zool. congol.* 5: 70–71.

———. 1929. Das Vorkommen des Schimpansen auf dem linken Kongo-Ufer. *Rev. Zool. Bot. afr.* 16: 425–26.

———. 1934. On the local races of the chimpanzee. *Ann. Mag. Nat. Hist.* 10(13): 576–83.

———. 1936. The Sterkfontein Ape. *Nature* 138: 969.

Scrimshaw, N. S., and V. R. Young. 1976. The requirements of human nutrition. *Sci. Am.* (in Japanese) 11: 33–47.

Seward, J. 1983. *Japanese in Action*. 2d rev. ed. New York: Weatherhill.

Shea, B. T. 1984. An allometric perspective on the morphological and evolutionary relationships between pygmy (*Pan paniscus*) and common (*Pan troglodytes*) chimpanzees. In: *The Pygmy Chimpanzee: Evolutionary Biology and Behavior*, pp. 89–130. R. L. Susman, ed. New York: Plenum.

Sibley, C. G., and J. E. Ahlquist. 1984. The phylogeny of the hominid primates, as indicated by DNA-DNA hybridization. *J. Mol. Evol.* 20: 2–15.

Simpson, G. G. 1963. The meaning of taxonomic statements. In: *Classification and Human Evolution*, pp. 1–31. S. L. Washburn, ed. London: Methuen.

Small, M. F. 1988. Female primate sexual behavior and conception. *Curr. Anthropol.* 29(1): 81–100.

Socha, W. W. 1984. Blood groups of pygmy and common chimpanzees: A comparative study. In: *The Pygmy Chimpanzee: Evolutionary Biology and Behavior*, pp. 13–41. R. L. Susman, ed. New York: Plenum Press.

Sugiyama, Y. 1965. On the social change of hanuman langur troop (*Presbytis entellus*). *Primates* 6: 381–418.

————. 1969. Social behavior of chimpanzees in the Budongo forest, Uganda. *Primates* 10: 197–225.

————. 1977. Budongo no mori no chimpanji: Sono shakai kouzou. In: *Chimpanji-ki*, pp. 473–542. J. Itani, ed. Tokyo: Kodansha.

————. 1979. Tool-using and -making behavior in wild chimpanzees at Bossou, Guinea. *Primates* 2(4): 513–24.

Sugiyama, Y., and J. Koman. 1979. Social structure and dynamics of wild chimpanzees at Bossou, Guinea. *Primates* 20: 323–39.

Susman, R. L., N. L. Badrian, and A. J. Badrian. 1980. Locomotor behavior of *Pan paniscus* in Zaire. *Am. J. Phys. Anthropol.* 53: 69–80.

Suzuki, A. 1971. Carnivority and cannibalism observed among forest-living chimpanzees. *J. Anthrop. Soc. Nippon* 79(1): 30–48.

————. 1977. Chimpanji no shakai to tekiou. In: *Chimpanji-ki*, pp. 251–336. J. Itani, ed. Tokyo: Kodansha.

Takahata, Y., T. Hasegawa, and T. Nishida. 1984. Chimpanzee predation in the Mahale Mountains from August 1979 to May 1982. *Int. J. Primat.* 5(3): 213–33.

Takenaka, O. 1983. Bunshi tokei ga shimesu eda-wakare no toki. [The evolutionary divergence timed by the molecular clock.] *Kagaku-asahi*, 1983(11): 39–41.

Taylor, H. M., ed. 1979. *English and Japanese in Contrast.* New York: Regents.

Teleki, G. 1973. *The Predatory Behavior of Wild Chimpanzees.* Lewisburg, PA: Bucknell University Press.

————. 1981. The omnivorous diet and eclectic feeding habits of chimpanzees in Gombe National Park, Tanzania. In: *Omnivorous Primates: Gathering and Hunting in Human Evolution*, pp. 303–43. R. S. O. Harding, and G. Teleki, eds. New York: Columbia University Press.

Tobias, P. V. 1965. Early man in East Africa. *Science* 149: 22–33.

Tratz, E., and H. Heck. 1954. Der afrikanische anthropoide "Bonobo," eine neue Menschenaffengattung. *Saugetierkd. Mitt.* 2: 97–101.

Tulp, N. 1641. *Observationum medicarum libri teres*, pp. 274–79. Amsterdam.

Turnbull, C. M. 1968. The importance of flux in two hunting societies. In: *Man the Hunter*, pp. 132–37. R. B. Lee, and I. DeVore, eds. Chicago: Aldine.

Tutin, C. E. G. 1979. Mating patterns and reproductive strategies in a community of wild chimpanzees (*Pan troglodytes schweinfurthii*). *Behav. Ecol. Sociobiol.* 6: 29–38.

Tutin, C. E. G., and P. R. McGinnis. 1981. Chimpanzee reproduction in the wild. In: *Reproductive Biology of the Great Apes: Comparative and Biomedical Perspectives*, pp. 239–64. C. E. Graham, ed. New York: Academic Press.

Tutin, C. E. G., and W. C. McGrew. 1973a. Chimpanzee copulatory behavior. *Folia primatol.* 19: 237–56.

————. 1973b. Sexual behaviour of group-living adolescent chimpanzees. *Am. J. Phys. Anthropol.* 38: 195–200.

Tuttle, R. H. 1969. Knuckle-walking and the problem of human evolution. *Science* 166: 953–61.

Tyson, E. 1699. *Orang-Outang, sive Homo Sylvestris: or, the Anatomy of a Pygmie Compared with That of a Monkey, and Ape, and a Man*. London: T. Bennett and Brown.

Urbain, A., and P. Rode. 1940. Un chimpanzé pygmée (*Pan satyrus paniscus* Schwarz) au Parc Zoologique du Bois de Vincennes. *Mammalia* 4: 10–12.

Van Maanen, J. 1988. *Tales of the Field: On Writing Ethnography*. Chicago: University of Chicago Press.

Vuanza, P. N., and M. Grabbe. 1975. *Les régimes moyens et extrêmes des climats principaux du Zaire*. Centre Meteorologique de Kinshasa.

Werner, O., and G. M. Schoepfle. 1988. *Systematic Fieldwork*, Vol. 1. Newbury, Calif.: SAGE.

White, T. D. 1983. The descendants of *Australopithecus afarensis*. Kagaku-asahi [translated interview], 1983(11): 27–31.

Wrangham, R. W. 1977. Feeding behaviour of chimpanzees in Gombe National Park, Tanzania. In: *Primate Ecology*, pp. 503–38. T. H. Clutton-Brock, ed. London: Academic Press.

———. 1979. Sex differences in chimpanzee dispersion. In: *The Great Apes: Perspectives on Human Evolution*, pp. 481–89. D. Hamburg, and E. McCown, eds. Menlo Park, CA: Benjamin/Cummings.

Yerkes, R. M. 1925. *Almost Human*. New York: Century.

———. 1939. Social dominance and sexual status in the chimpanzee. *Hum. Biol.* 11: 78–111.

Yerkes, R. M., and J. H. Elder. 1936. Oestrus, receptivity, and mating in the chimpanzee. *Comp. psychol. Monogr.* 13: 1–39.

Yerkes, R. M., and B. W. Learned. 1925. *Chimpanzee Intelligence and Its Vocal Expressions*. Baltimore.

Yerkes, R. M., and A. W. Yerkes. 1929. *The Great Apes: A Study of Anthropoid Life*. New Haven: Yale University Press.

Young, W. C., and R. M. Yerkes. 1943. Factors influencing the reproductive cycle in the chimpanzee: The period of adolescent sterility and related problems. *Endocrinology* 33: 121–54.

Zihlman, A. L. 1982. *The Human Evolution Coloring Book*. New York: Barnes and Noble.

———. 1984. Body build and tissue composition in *Pan paniscus* and *Pan troglodytes*, with comparisons to other hominoids. In: *The Pygmy Chimpanzee: Evolutionary Biology and Behavior*, pp. 179–200. R. L. Susman, ed. New York: Plenum.

Zihlman, A. L., and D. L. Cramer. 1978. Skeletal differences between pygmy (*Pan paniscus*) and common chimpanzees (*Pan troglodytes*). *Folia primatol.* 29: 86–94.

Zihlman, A. L., J. E. Cronin, D. L. Cramer, and V. M. Sarich. 1978. Pygmy chimpanzee as a possible prototype for the common ancestor of humans, chimpanzees and gorillas. *Nature* 275: 744–46.

Zuckerman, S. 1932. *The Social Life of Monkeys and Apes*. London: Kegan Paul, Trench & Trubner.

Index

In this index an "f" after a number indicates a separate reference on the next page, and an "ff" indicates separate references on the next two pages. A continuous discussion over two or more pages is indicated by a span of page numbers, e.g., "57–59." *Passim* is used for a cluster of references in close but not consecutive sequence.

Library of Congress Cataloging-in-Publication Data

Kanō, Takayoshi, 1938-
  [Saigo no ruijin 'en. English]
  The last ape : pygmy chimpanzee behavior and ecology / by
Takayoshi Kano ; translated by Evelyn Ono Vineberg.
    p.   cm.
  Translation of: Saigo no ruijin 'en.
  Includes bibliographical references and index.
  ISBN 0-8047-1612-9 :
  1. Pygmy chimpanzee—Behavior.   2. Pygmy chimpanzee—Ecology.
QL737.P96K27   1992
599.88'44—dc20   91-2147
                                                          CIP
                                                          rev.

♾ This book is printed on acid-free paper.